5G Backhaul and Fronthaul

5G Backhaul and Fronthaul

Edited by

Esa Markus Metsälä
Nokia Networks, Finland

Juha T. T. Salmelin
Nokia Networks, Finland

Registered Office
John Wiley & Sons, Inc., 111 River Street, Hoboken, NJ 07030, USA

Editorial Office
John Wiley & Sons Ltd, The Atrium, Southern Gate, Chichester, West Sussex, PO19 8SQ, UK

For details of our global editorial offices, customer services, and more information about Wiley products visit us at www.wiley.com.

Wiley also publishes its books in a variety of electronic formats and by print-on-demand. Some content that appears in standard print versions of this book may not be available in other formats.

Library of Congress Cataloging-in-Publication Data
Names: Metsälä, Esa, editor. | Salmelin, Juha, editor.
Title: 5G backhaul and fronthaul / Esa Markus Metsälä, Juha T. T. Salmelin.
Description: Hoboken, NJ: John Wiley & Sons, 2023. | Includes bibliographical references and index.
Identifiers: LCCN 2022054499 (print) | LCCN 2022054500 (ebook) | ISBN 9781119275640 (hardback) | ISBN 9781119275664 (pdf) | ISBN 9781119275572 (epub) | ISBN 9781119275671 (ebook)
Subjects: LCSH: 5G mobile communication systems.
Classification: LCC TK5103.25 .A115 2023 (print) | LCC TK5103.25 (ebook) | DDC 621.3845/6–dc23/eng/20230111
LC record available at https://lccn.loc.gov/2022054499
LC ebook record available at https://lccn.loc.gov/2022054500

Cover Design: Wiley
Cover Image: © metamorworks/Shutterstock

Set in 9.5/12.5pt STIXTwoText by Integra Software Services Pvt. Ltd, Pondicherry, India
Printed and bound by CPI Group (UK) Ltd, Croydon, CR0 4YY

C9781119275640_010323

Contents

Acknowledgements

The editors would first like to acknowledge all of the contributing authors, individually and also as a team: our thanks go to Mika Aalto, Pascal Dom, Akash Dutta, Kenneth Y. Ho, Harri Holma, Lieven Levrau, Raija Lilius, Esa Malkamäki, Antti Pietiläinen, Paolo di Prisco, Derrick Remedios, Andy Sutton and Antti Toskala.

For specific review efforts and large and small suggestions and contributions, we would like to thank our colleagues at Nokia, including Hannu Flinck, Markus Isomäki, Qinglong Qiu, Litao Ru, Olli Salmela, Peter Skov and Jeroen Wigard for specific input, in addition to a number of experts who have contributed during the preparation and review of the material. We are also grateful to the community of radio and networking professionals in the industry, with whom we have had the pleasure of discussing the topics covered in this book.

We thank the excellent team at John Wiley & Sons for very good co-operation and support during the time of writing.

We appreciate the patience and support of our families and our authors' families during the writing period and are grateful for this.

We welcome comments and suggestions for improvements or changes that could be implemented in forthcoming editions of this book. Feedback is welcome to the editors' email addresses: esa.metsala@nokia.com and juha.salmelin@nokia.com.

Esa Metsälä and Juha Salmelin
Espoo, Finland

About the Editors

Esa Metsälä leads the 5G Radio Access System Specification Team for networking at Nokia. He has extensive experience in leading mobile backhaul and mobile network system-level topics and teams since 2G.

Juha Salmelin manages Smart City initiatives for Nokia Networks. He has over 20 years' experience in mobile backhaul research as a team leader, department manager and technology head at Nokia.

List of Contributors

Mika Aalto
Nokia Cloud and Network Services
Espoo, Finland

Pascal Dom
Nokia Network Infrastructure
Antwerp, Belgium

Akash Dutta
Nokia Networks
Bengaluru, India

Kenneth Y. Ho
Nokia Networks (Retired)
New Jersey, USA

Harri Holma
Nokia Strategy& Technology
Espoo, Finland

Lieven Levrau
Nokia Network Infrastructure
Valbonne, France

Raija Lilius
Nokia Networks
Espoo, Finland

Esa Malkamäki
Nokia Strategy& Technology
Espoo, Finland

Esa Metsälä
Nokia Networks
Espoo, Finland

Antti Pietiläinen
Nokia Networks
Espoo, Finland

Paolo di Prisco
Nokia Networks
Milan, Italy

Derrick Remedios
Nokia Networks Infrastructure
Ottawa, Canada

Juha Salmelin
Nokia Networks
Espoo, Finland

Andy Sutton
British Telecom
Warrington, UK

Antti Toskala
Nokia Strategy& Technology
Espoo, Finland

1

Introduction

Esa Metsälä and Juha Salmelin

1.1   Introducing 5G in Transport

Mobile networks evolve, with new system generations bringing higher speeds and new service capabilities, with improved system architectures and novel radio technologies. With this evolution, 5G mobile backhaul and fronthaul play an increasing role, with new challenges in matching the new 5G system capabilities.

With 5G, higher peak rates are surely an important enhancement, as was the case with 4G. This is just a single item, however. 5G is the most versatile mobile system so far, supporting not only traditional mobile broadband but also new industrial and enterprise use cases with building blocks like network slicing for unique services.

URLLC services extend the 5G capability to applications which were previously not possible in a mobile system. The 5G network is built to serve different use cases and customers, and these impact transport also.

New high-frequency bands mean not only high capacity, but also a potentially huge amount of small cells. Further, radio signals from outdoor sites are heavily attenuated indoors. So, indoor solutions will be required for coverage with related connectivity solutions.

Disaggregation of the 5G radio access and related radio clouds opens up a new type of network implementation, where existing server and computing platforms can be leveraged to support virtualized or containerized network functions. At the same time, opening of the interfaces between network elements allows multi-vendor deployment and operation. With 5G radio energy efficiency is improved and network level optimizations allow further energy savings.

Packet-based fronthaul is a genuine new area, which evolved from previous industry proprietary solutions to use common packet networking technologies, which in turn better allow shared packet infrastructure in fronthaul also, but with special emphasis on the time sensitivity of fronthaul traffic flows and synchronization.

This all shows the range of new tasks there are for 5G mobile backhaul and fronthaul, since there is no longer a single use case or single implementation approach for radio networks.

As transport is all about connecting mobile network elements in an efficient way, there will be many different needs based on the radio cloud and the level of disaggregation targeted, and based on services intended to be delivered over the system, and these all have to be supported on the very same backhaul and fronthaul network.

For example, essential topics like transport latency and capacity requirements all depend on the use cases as well as the type of disaggregation and virtualization deployed.

The focus areas of 5G backhaul and fronthaul are:

- Fronthaul, which in 5G is a new packet-based latency critical interface that – with eCPRI and O-RAN – has evolved from previous CPRI-based interfaces.
- Radio cloud, which has connectivity needs within and through the data centre to other sites, with related virtualization or containerization of network functions, and needs for scaling and flexible connectivity.
- Accurate synchronization support, due to the 5G TDD radio.
- Matching the increased 5G system performance targets in transport, with reduced latency and higher reliability and availability.
- Making the backhaul and fronthaul secure and robust against both intended and unintended anomalies with cryptographic protection.
- Introducing new 5G services through backhaul and fronthaul with ultra reliability and low latency and supporting new use cases – like industrial and enterprises – over backhaul and fronthaul.
- Future proofness and adaptability for different deployments and support for the cloud-native 5G network.
- The trend for more openness in network management, in object modelling and in related interfaces, driven by O-RAN and other forums.
- Making the network programmable to speed up service creation and enable managing a network that encompasses huge amounts of RUs as well as new millimetre-wave small cells.
- Close cooperation with LTE for dual connectivity, NSA mode, reuse of sites and transport network, with continued support for legacy protocols and technologies.

Transport links are to be upgraded for 5G. Physical media are fibre complemented, with new high-bandwidth wireless links that offer 10G and more capacity.

Network connectivity services focus on resiliency, flexibility, security and future proofness for network expansion and new services. All network resources have to be protected against unauthorized use and communications in the network domain have to be secure.

Synchronization is of high importance since the TDD mode of the 5G system requires phase and time synchronization. Many cases rely on the transport network to deliver synchronization.

In transport networks, major planning cycles occur once in a while – around every 10 years – and at these moments, physical network topologies and physical media usage are also modified. Connectivity layer upgrade is a slightly smaller task once the physical media is in place, since network equipment on existing sites can be changed more easily and with less effort than laying new fibre, even though equipment changes involve not only capex costs but also (more importantly) related training and competence development costs for the new technologies.

For mobile backhaul and fronthaul, the introduction of the 5G system is a trigger for an upgrade cycle – at minimum, capacity additions in links and nodes, and possibly including enhancements and modernization of the network in a wider scope which can include in addition to new networking functionality also retiring links and nodes that are poorly performing in terms of bit errors, packet loss, latency, energy efficiency or security.

1.2 Targets of the Book

While there are multiple challenges identified above for 5G backhaul and fronthaul, there are also new technologies and improvements that address such challenges.

This book aims to cover the key requirements for 5G mobile backhaul and fronthaul and also open up some of the inner workings of the 5G system, especially the 5G radio access network, in order to find the transport services that are most feasible.

Once the drivers for the networking layers are covered, the book concentrates on different technologies that address the needs 5G brings. Some specific use cases and configurations are discussed, with the intent of summarizing the technical aspects covered.

Chapter 2 introduces the 5G system and Chapter 3 follows on in analysing key technical and economic considerations of 5G backhaul and fronthaul, from an operator viewpoint. Next, Chapter 4 discusses the requirements for 5G backhaul and fronthaul performance and other drivers.

Chapter 5 focuses on essential new 5G items related to transport, like network slicing, IAB (Integrated Access and Backhaul), NTNs (Non-Terrestrial Networks) and private networks for industry solutions and shared networks for smart cities.

Chapter 6 covers physical layer technologies in the optical domain. The optical domain covers a wide range of technologies, from passive to active WDM and also PONs (Passive Optical Networks). Fibre is the medium that provides ample capacity for 5G backhaul and fronthaul.

Optical, while often the preferred choice, needs to be complemented with wireless links – especially in access. Going from traditional microwave bands to E- and D-bands gives capacities of 10G, 25G and more, which means that wireless continues to remain a feasible option for many sites and can bridge the gap in those areas where fibre deployment is not economical. This is the topic of Chapter 7. In rare cases, optical wireless links can also be considered. These FSOs (Free Space Optics) are presented at the end of the chapter.

Chapter 8 continues with key cloud and networking technologies. Cloud nativeness, when applied to 5G radio, is not a trivial topic itself and for networking it introduces new concepts and requirements. New technologies – like segment routing – can help for a light version of traffic engineering, while fronthaul requires time-sensitive networking solutions.

In Chapter 9, example deployment cases are presented, covering – in addition to traditional macro deployment for 5G – also cloud, fronthaul, industrial, indoor and synchronization cases. The book concludes with a summary in Chapter 10.

1.3 Backhaul and Fronthaul Scope within the 5G System

In 5G service delivery, transport on the backhaul and fronthaul is an integral and important part of the 5G system infrastructure.

The 5G mobile network elements in both the 5G core and 5G radio include IP endpoints for communication with peer(s); then they interface the backhaul or fronthaul networks, which operate as their own networking domain, with physical layer facilities like fibre cabling and networking devices that are separate from the mobile network elements, with their own network management.

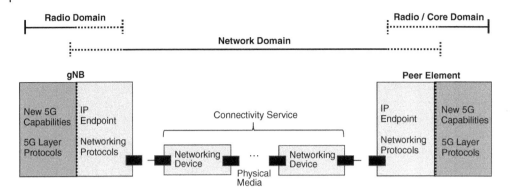

Figure 1.1 Network domain within the 5G system.

Figure 1.1 shows the gNB communicating with a peer element, which could be another 5G RAN element or a 5G core network element. Networking devices and physical transport media provide the required connectivity, which in many cases is useful to abstract as a connectivity service.

The IP endpoints of gNB, related processing and IP interfaces are somewhat in a grey area between radio and the network domain: they are part of the radio domain in the sense that they form part of the base station. They are also part of the network domain since the protocols processed are networking protocols not radio protocols.

Often, the 5G radio elements (including IP endpoints) are administered by the radio team and the external networking devices by the transport team. In the network domain, devices may further be operated by dedicated teams based on competence (e.g. the optical layer is managed by one competence pool, wireless transmission by another and the IP layer by yet another expert team).

In this book, 'transport' is used interchangeably with 'networking' to cover protocols from the transport physical layer and media up to the IP layer but touching also protocols on top of the IP layer, like UDP and SCTP. The 5G layer protocols include, for example, GTP-U and application protocols like NG-AP and all 5G RAN internal radio protocols like PDCP, RLC, MAC and radio L1.

Backhaul refers to the RAN–core network interface (NG2, NG3), midhaul to the RAN internal F1 interface (which is the high-level fronthaul split point defined by 3GPP) and fronthaul to the RAN internal low-level split point (which is defined by O-RAN).

1.4 Arranging Connectivity within the 5G System

The connectivity service is supplied in-house by a transport team or a third-party connectivity service is used. Often, a combination of the two – some network tiers or geographical areas are served by own transport assets, while others rely on external service providers.

Service Level Agreements (SLAs) describe essential characteristics of the service at the User-to-Network Interface (UNI). MEF (Metro Ethernet Forum) is an organization focusing on standardized service definitions that are useful for 5G backhaul and fronthaul.

The connectivity service concept includes CE (Customer Equipment) and PE (Provider Edge) – CE meaning equipment on the customer side and PE the peer element on the service provider side. This terminology is used, for example, with IP VPNs.

When the service is abstracted, the user of the service is agnostic to how the actual network is implemented and what is the current operating state of the underlying individual network links and nodes. This simplifies many topics for the user as there is no need to worry about the details in implementation.

The implementation aspects sometimes become visible, e.g. when the underlying network path changes, causing change in the experienced latency, or when there are network anomalies like bit errors. As long as these are within the SLA of the service, the user should not be concerned. What is of importance is the possibility to monitor that the actual obtained service level matches the agreed SLA which the user is paying for.

With packet-level service, there are many possible service-impacting topics to be considered. Ultimately, a complete fibre can be leased, in which case the user has a hard guarantee and can decide to use full bandwidth without interference from other traffic sources, but this is not feasible for all cases and fibre bandwidth can be shared by wavelengths or by packet (IP) nodes.

On the implementation side for backhaul and fronthaul, wireless devices, photonic devices and switches, routers and security gateways are seldom all sourced from the same vendor. Multi-vendor interoperation is addressed by devices complying with relevant standards and interoperability testing is needed to guarantee correct operation. Simplification of the solution can lead to significant benefits but is not trivial to achieve. Open interfaces and the approach towards a programmable network are key for network integration, efficient operation and intended performance.

Acquiring sites and rights of way for fibre instalments is costly and time-consuming, and one possibility is that transport – especially physical resources like fibre but possibly also higher-layer operations – is shared in such a way that one entity owns and operates the network. This entity can be a neutral host, meaning a third-party 'independent entity' offers services to multiple tenants.

All these aspects impact 5G backhaul and fronthaul, which is in practice multi-technology and multi-vendor, with multiple feasible operating models.

1.5 Standardization Environment

The standardization environment is heterogenous regarding 5G backhaul and fronthaul. Since only key protocol definitions are referred to in 3GPP, there are a wide range of networking-related standard defining organizations and industry forums that define technical areas and individual topics related to 5G backhaul and fronthaul. Some of these are briefly introduced here.

1.5.1 3GPP and other organizations

In 5G as in previous mobile generations, transport is standardized in 3GPP for the logical interface protocol stacks [1–9]. 3GPP reuses and refers to existing, primarily IETF standards, where networking protocols are defined. The layers below IP in 3GPP are referred to as the Data Link Layer and the Physical Layer, so are not defined in the 3GPP protocol stacks. In practice, Ethernet is typically used for L2/L1 over fibre or wireless media.

The borderline between 3GPP and networking standards is UDP/IP in the user plane and SCTP/IP in the control plane. 3GPP defines the user-plane GTP-U tunnelling protocol below which the UDP/IP layer is defined in IETF RFCs. GTP-U tunnelling is defined in Ref. [10]. In the control plane, 5G application protocols such as NG-AP are defined in 3GPP while the lower layer, SCTP, is defined in IETF.

5G system architecture includes network elements like gNB and then logical interfaces between the elements, defining which element needs to have connectivity to which other elements. For the actual backhaul and fronthaul network implementation, 3GPP is agnostic.

Connectivity may be arranged with any technology that enables communication between the IP endpoints of the 5G elements (e.g. between gNB and the UPF). The design of the communication service and its characteristics are left for network implementation.

Security is covered for the network domain and 3GPP mandates the use of cryptographic protection in the network domain, with IP security suite of protocols in 3GPP TS 33.210 [11]. Implementations differ also with IPsec e.g. whether the IPsec endpoints reside on the mobile elements themselves or in a separate security gateway [SEG]. Additionally, DTLS is included in the control plane logical interface definitions of 5G for 38.412 NG Signalling Transport, 38.422 Xn Signalling Transport and 38.472 F1 Signalling Transport [4, 6, 8].

Assigning 5G functions into the network elements, delivery of 5G services at the intended user experience and operation of the 5G radio protocols impose requirements for many backhaul and fronthaul aspects, such as maximum delay or availability of the connectivity service. These requirements are, however, also dependent on the network implementation and mobile network operator's targets, and so a single, generic value cannot be found.

The synchronization method is likewise an implementation topic – how is accurate timing supplied? The accuracy requirement itself originates from the 5G air interface and 5G features. ITU-T and IEEE have defined profiles for synchronization using the IEEE 1588 protocol, and O-RAN also covers the topic.

Fronthaul, or the low-layer split point, is not defined by 3GPP in detail but by O-RAN, so it is a special case compared to the 3GPP logical interface definitions.

IETF: Internet Engineering Task Force

As 3GPP reuses existing networking standards, the role of IETF is important as the protocols for the IP layer – including IP security – originate from there. All logical interfaces rely on IPv6 or IPv4 and IPsec protocols, with commonly Ethernet L2/L1 beneath the IP protocol layer. The IP protocols are defined by IETF RFCs as well as UDP, SCTP, DTLS and IPsec [12–21].

IETF definitions for services like IP and Ethernet VPNs (RFC 4364 BGP/MPLS IP Virtual Private Networks [VPNs] IP VPN and RFC 7432 BGP/MPLS IP Virtual Private Networks [VPNs] Ethernet VPN) are examples of implementation options for the transport network [22, 23]. With work progressing in IETF for new services and capabilities, they will become available also for implementation in mobile backhaul and fronthaul. Some examples of work in progress in IETF relate to segment routing and deterministic networks [24, 25]. IETF is also source for many Yang data models.

With IP networking, the basic IP networking protocols are relevant, like the companion protocols ARP (RFC 862) and ICMP (RFC 792, RFC 4443) [26–28], for example, routing protocols, monitoring and OAM capabilities, and many other definitions.

IEEE: Institute of Electrical and Electronics Engineers

IEEE is the home for Ethernet standards which are in use in both backhaul and fronthaul as de-facto physical interfaces and additionally in other applications.

IEEE 802.3 works on Ethernet physical layer specifications. For 5G fronthaul and backhaul, application data rates of 10G and more are of most interest, like 25G, 50G and 100G.

10G is standardized as IEEE 802.3ae, including both short-range multimode (-SR) and long-range singlemode (-LR) fibre media. For 25G, 25GBASE-SR is standardized as 802.3by and 25GBASE-LR as 802.3cc. These are included in the 2018 update of 802.3 [29].

The single-lane rate for 50 Gbps is in IEEE 802.3cd [30], using PAM-4 modulation. The standard for single-lane 100 Gbps is ongoing by IEEE 802.3ck [31].

Ethernet bridging is specified in 802.1D-2004. This standard has been amended (e.g. with 802.1Q [VLANs], 802.1ad [Provider Bridges], 802.1ah [Provider Backbone Bridges] and 802.1ag [Connectivity Fault Management]). These and other amendments are included in the 2018 update of IEEE 802.1Q [32].

With the Ethernet, cryptographic protection has been added with 802.1X (802.1X-2010 Port-based Network Access Control), 802.1AE (802.1AE-2006 MAC Security) and 802.1AR (Secure Device Identity) [33–35].

An active working area related to 5G fronthaul is Time-Sensitive Networking (TSN). Multiple TSN amendments to the base standard (802.1Q) – also touching 802.3 – have already been included or are in progress. A separate profile for 5G fronthaul is defined in IEEE 802.1CM [36]. Another fronthaul-related standard is for carrying CPRI over Ethernet, in IEEE P1914.3 [37].

The precision timing protocol (IEEE 1588v2) [38] is used for delivering synchronization for the base stations, including phase and time synchronization for 5G.

O-RAN: Open RAN Alliance

O-RAN [39] defines a fronthaul standard that is derived from the eCPRI, with related protocol stacks and functionality.

The O-RAN fronthaul interface is defined in Ref. [40]. O-RAN also has other standards with networking impacts, related to cloud and disaggregated RAN and definitions for RIC. Another important area is data/object models that are defined using Yang, enabling openness in network management.

ITU-T: International Telecommunications Union

ITU-T [41] covers a wide range of transmission-, system- and synchronization-related standards. Many relevant items include optical and photonics standards and profile definitions of timing distribution using the precision timing protocol (IEEE 1588).

ISO: International Organization for Standardization

ISO, in the networking area, is known for the OSI (Open System Interconnection) model and is the origin of the IS-IS routing protocol definition [42]. IS-IS is included in the IEEE work for TSN, which makes the protocol potentially relevant in the 5G fronthaul area.

MEF: MetroEthernet Forum

MEF is the source for service definitions for Ethernet services, with dedicated definitions for backhaul and fronthaul [43]. Service definitions have been expanded to cover IP services also.

Related to the Ethernet rates and transceivers, multiple industry forums are relevant. The Ethernet Technology Consortium [44] and the Optical Internetworking Forum (OIF) [45] have been active in driving new Ethernet MAC rates. Many of these initiatives have since resulted in an IEEE standard.

The SFF (Small Form Factor) committee is similarly an industry forum (now Storage Networking Industry Association [SNIA] Technology Affiliate Technical Working Group) [46] for optical pluggables.

NGMN Alliance (Next Generation Mobile Networks Alliance) is an open forum created by network operators to help the industry with clear operator requirements and guidance for better mobile networks [47]. Many topics cover or impact also 5G backhaul and fronthaul.

ETSI (European Telecommunications Standards Institute) [48] has worked among many other topics (e.g. for network function virtualization definitions and a virtualization framework covering also management and orchestration).

An industry foundation, CNCF (Cloud Native Computing Foundation) [49], covers topics of cloud native computing, container management and orchestration.

The CPRI forum [50] is an industry cooperation that has defined interfaces between the RE (Radio Equipment) and REC (Radio Equipment Control), including CPRI and eCPRI. Recently, new work on fronthaul has been in progress within O-RAN.

References

1 3GPP TS 36.414 S1 Data Transport.

2 3GPP TS 36.422 X2 Signalling Transport.

3 3GPP TS 36.424 X2 Data Transport.

4 3GPP TS 38.412 NG Signalling Transport.

5 3GPP TS 38.414 NG Data Transport.

6 3GPP TS 38.422 Xn Signalling Transport.

7 3GPP TS 38.424 Xn Data Transport.

8 3GPP TS 38.472 F1 Signalling Transport.

9 3GPP TS 38.474 F1 Data Transport.

10 3GPP TS 29.281 General Packet Radio System (GPRS) Tunnelling Protocol User Plane (GTPv1-U).

11 3GPP TS 33.210 Network Domain Security (NDS); IP network layer security.

12 IETF RFC 768 User Datagram Protocol.

13 IETF RFC 8200 Internet Protocol, Version 6 (IPv6) Specification.

14 IETF RFC 791 Internet Protocol.

15 IETF RFC 2474 Definition of the Differentiated Services Field (DS Field) in the Ipv4 and Ipv6 Headers.

16 IETF RFC 4303 IP Encapsulating Security Payload (ESP).

17 IETF RFC 4301 Security Architecture for the Internet Protocol.

18 IETF RFC 6311 Protocol Support for High Availability of IKEv2/IPsec.

19 IETF RFC 7296 Internet Key Exchange Protocol Version 2 (IKEv2).

20 IETF RFC 4960 Stream Control Transmission Protocol.

21 IETF RFC 6083 Datagram Transport Layer Security (DTLS) for Stream Control Transmission Protocol (SCTP).

22 IETF RFC 4364 BGP/MPLS IP Virtual Private Networks (VPNs) IP VPN.

23 IETF RFC 7432 BGP/MPLS IP Virtual Private Networks (VPNs) Ethernet VPN.

24 https://datatracker.ietf.org/wg/spring/documents.

25 https://datatracker.ietf.org/wg/detnet/about.

26 IETF RFC 862 An Ethernet Address Resolution Protocol.

27 IETF RFC 792 Internet Control Message Protocol.

28 IETF RFC 4443 Internet Control Message Protocol (ICMPv6) for the Internet Protocol Version 6 (IPv6) Specification.

29 IEEE 802.3-2018 IEEE Standard for Ethernet.

30 IEEE802.3cd Media Access Control Parameters for 50 Gb/s and Physical Layers and Management Parameters for 50 Gb/s, 100 Gb/s, and 200 Gb/s Operation.

31 IEEE802.3ck Media Access Control Parameters for 50 Gb/s and Physical Layers and Management Parameters for 50 Gb/s, 100 Gb/s, and 200 Gb/s Operation.

32 IEEE 802.1Q - 2018 Bridges and Bridged Networks.

33 IEEE 802.1X-2010 Port-based Network Access Control.

34 IEEE 802.1AE MAC Security.

35 IEEE 802.1AR Secure Device Identity.

36 IEEE802.1CM Time-Sensitive Networking for Fronthaul.

37 IEEE P1914.3. Standard for Radio Over Ethernet Encapsulations and Mappings.

38 IEEE 1588-2019 Standard for a Precision Clock Synchronization Protocol for Networked Measurement and Control System.

39 www.o-ran.org.

40 O-RAN.WG4.CUS.0-v07.00, Control, User and Synchronization Plane Specification.

41 www.itu.int/en/ITU-T/Pages/default.aspx.

42 www.iso.org/home.html.

43 www.mef.net.

44 ethernettechnologyconsortium.org.

45 www.oiforum.com.

46 www.snia.org.

47 https://www.ngmn.org.

48 www.etsi.org.

49 www.cncf.io.

50 http://www.cpri.info.

2

5G System Design Targets and Main Technologies

Harri Holma and Antti Toskala

2.1   5G System Target

5G radio represents a major step in mobile network capabilities. So far, mobile networks have mainly provided connectivity for smartphones, tablets and laptops. 5G will take traditional mobile broadband to the extreme in terms of data rates, capacity and availability. Additionally, 5G will enable further capabilities including massive Internet of Things (IoT) connectivity and critical communication. 5G targets are illustrated in Figure 2.1. 5G is not only about new radio or new architecture or core, but also about the number of new use cases. It is expected that 5G will fundamentally impact the whole of society in terms of improving efficiency, productivity and safety. 4G networks were designed and developed mainly by telecom operators and vendors for the smartphone use case. There is a lot more interest in 5G networks by other parties, including different industries and cities, to understand 5G capabilities and push 5G availability. 5G is about connecting everything in the future.

5G radio can bring major benefits in terms of network performance and efficiency, see summary in Figure 2.2. We expect substantially higher data rates – up to 20 Gbps, clearly lower cost per bit, higher spectral efficiency, higher network energy efficiency and lower latency. The values are based on the following assumptions:

- The shortest transmission time in the 5G layer is <0.1 ms, which enables a round-trip time of 1 ms.
- Energy efficiency assumes a three-sector 100 MHz macro base station busy hour average throughput of 1 Gbps, busy hour share of 7% and base station average power consumption of 2 kW. The efficiency improvement of 10× compared to LTE is obtained with power saving techniques at low load and wideband carrier up to 100 MHz in the 3.5 GHz band.
- A peak rate of 1.5 Gbps assumes 100 GHz bandwidth TDD with 4×4MIMO and 256QAM modulation and a peak rate of 5 Gbps assumes 800 MHz bandwidth TDD with 2×2MIMO.
- Spectral efficiency of 10 bps/Hz/cell assumes the use of massive MIMO (Multiple Input Multiple Output) beamforming and four antenna devices. The typical LTE downlink efficiency is 1.5–2.0 bps/Hz/cell in the live networks and +50% more with 4×4MIMO.

5G Backhaul and Fronthaul, First Edition. Edited by Esa Markus Metsälä and Juha T. T. Salmelin.

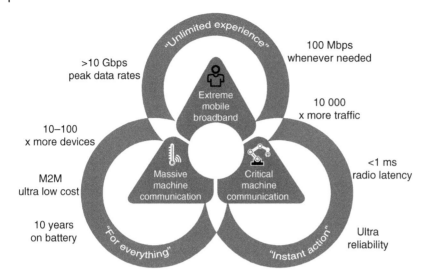

Figure 2.1 5G targets.

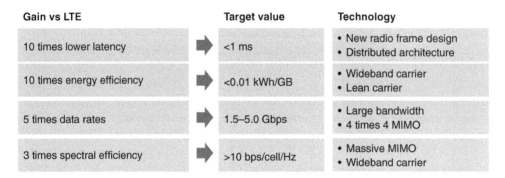

Figure 2.2 Summary of 5G technology capabilities.

2.2 5G Technology Components

The high targets of 5G networks require a number of new technologies. The main new technology components are shown in Figure 2.3.

1) New spectrum. The very high data rates (up to 5–10 Gbps) require bandwidth up to 800 MHz, which is available at higher frequency bands. 5G is the first radio technology designed to operate on any frequency bands between 450 MHz and 90 GHz. The low bands are needed for good coverage and the high bands for high data rates and capacity. The frequencies above 30 GHz have wavelength <1 cm and are commonly called millimetre waves (mmWs). Sometimes also lower frequencies (24–28 GHz) are included in the mmW notation. LTE specifications are not defined beyond 6 GHz.

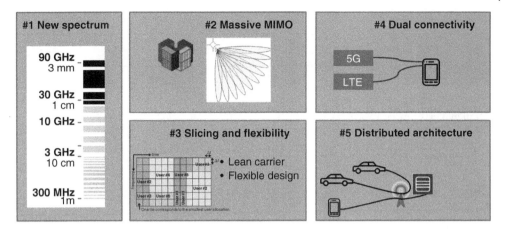

Figure 2.3 Key 5G technology components.

2) Massive MIMO with beamforming increases spectral efficiency and network coverage substantially. Beamforming becomes more practical at higher frequencies because the antenna size is relative to the wavelength. In practice, massive MIMO can be utilized at frequencies above 2 GHz in the base stations and at mmWs even in the devices. Massive MIMO will be part of 5G from day 1, which will avoid any legacy device problems. User-specific beamforming was not supported in the first LTE release but added only later in 3GPP.

3) Flexible air interface design and network slicing. Physical and protocol layers in 5G need a flexible design in order to support different use cases, vertical segments, different frequency bands and to maximize the energy and spectral efficiency. Network slicing will create virtual networks for the different use cases within the same 5G network.

4) Dual connectivity. 5G can be deployed as a standalone system but more typically 5G will be deployed together with LTE. The 5G device can have simultaneous connection to 5G and LTE. Dual connectivity can increase the user data rate and improve the connection reliability.

5) Distributed architecture with cloud flexibility. The typical architecture in LTE networks is fully distributed in the radio and fully centralized in the core network. Low latency requires bringing the content close to the radio network, which leads to local breakout and Multi-access Edge Computing (MEC). Scalability requires bringing cloud benefits to radio networks with an edge cloud and local cloud architecture.

3GPP completed the first version of 5G specifications in December 2017, with specification freeze and stability of specifications for the NSA (Non-Standalone Architecture) reached by end 2018. We list the main decisions below:

- Waveform. The downlink solution is an OFDM (Orthogonal Frequency Division Multiplexing)-based waveform with Cyclic Prefix (CP), which is similar to the LTE waveform. The uplink waveform is also an OFDM with single carrier (SC-FDMA/DFT-S-OFDM) option for coverage extension. The uplink solution is different from that in LTE, where only a single carrier is used in the uplink. Having similar downlink and

Table 2.1 5G numerology in 3GPP Release 15.

Subcarrier spacing [kHz]	15	30	60	120	240**
Symbol duration [µs]	66.7	33.3	16.6	8.33	4.17
Nominal CP [µs]	4.7	2.41	1.205	0.60	0.30
Nominal max carrier BW [MHz]	50	100	200	400	—
Max FFT size	4096	4096	4096	4096	—
Min scheduling interval (symbols)	14	14	14	14	—
Min scheduling interval (slots)*	1	1	1	1	—
Min scheduling interval (ms)	1.0	0.5	0.25	0.125	—

*2/4/7 symbol mini-slot for low-latency scheduling.
**SS block only.

uplink solutions in 5G simplifies beamforming with reciprocal channel, interference avoidance, device-to-device communication (side-link) and in-band backhauling. There is a 1–2 dB coverage penalty in OFDM compared to single carrier, and therefore a single-carrier option is needed also in the 5G uplink.

- Channel coding. Data channels use Low Density Parity Check (LDPC) coding and control channels use Polar coding or Reed–Muller coding depending on the number of bits. Both coding solutions are different than in LTE where data channels use Turbo coding and control channels use convolutional and Reed–Muller coding. The main reason for using LDPC is the decoding complexity: clearly less silicon area is required for the decoding process with LDPC compared to Turbo decoding. LDPC is used also in the latest IEEE Wi-Fi specifications but there are some differences between 3GPP and IEEE usage of LPDC. Control channel coding solution was less critical due to the small data rate of the control channel.
- The 3GPP numerology is shown in Table 2.1. 5G is designed to support a number of subcarrier spacing and scheduling intervals, depending on the bandwidth and latency requirements. Subcarrier spacings between 15 kHz and 120 kHz are defined in Release 15 for data channels and 240 kHz for Synchronization Signal (SS). The narrow subcarrier spacings are used with narrow 5G bandwidths and are better for extreme coverage. If we consider a typical 5G deployment in the 3.5 GHz band, the bandwidth could be 40–100 MHz and the subcarrier spacing 30 kHz. The corresponding numbers in LTE are 20 MHz bandwidth and 15 kHz subcarrier spacing. 5G subcarrier spacing is designed to be 2^N multiples of 15 kHz. If the slot length is >0.125 ms in the narrowband cases and low latency is required, then a so-called mini-slot can be used where the transmission time is shorter than one slot. It is also possible to combine multiple slots together.

2.3 Network Architecture

The overall 5G system architecture has been reworked in line with the following key principles:

- Stateless core network entities
- Separate control and data handling
- Support for both data-centric and service-based architecture
- Support for centralized data storage
- Support of local and centralized service provision
- Minimized access and core network dependencies

The overall architecture is shown in Figure 2.4, illustrating the separate control and user plane elements as well as separate data storage in the network.

The control plane has the following main elements:

- Access and Mobility Management Function (AMF), covering mobility management functionality on the core network side as well as being the termination for the RAN control plane interface to the core (NG2 in Figure 2.5). Also, authentication and authorization are taken care of by the AMF and, respectively, the NAS terminates in the AMF.
- Session Management Function (SMF), covering roaming functionality, UE IP address allocation and management, as well as selection and control of the user plane function. The SMF is also the termination point towards policy control and charging functions.
- Policy Control Function (PCF), providing policy rules for the control plane functions. The PCF gets subscriber information from the data storage functions.
- Local and centralized gateways – the User Plane Functions (UPFs) – which are the connection anchor points in connection with mobility. The UPFs cover packet routing and forwarding and take care of the QoS handling for user data. Possible packet inspection and policy rule enforcements are also covered by the UPFs.

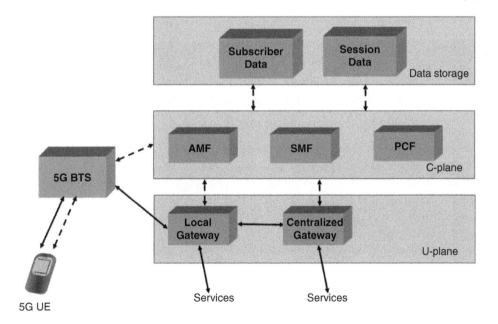

Figure 2.4 Overall 5G network architecture.

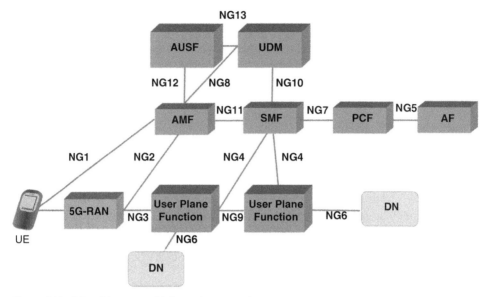

Figure 2.5 5G architecture with key reference points.

The main reference points in the 5G architecture are shown in Figure 2.5, with Unified Data Management (UDM) taking care of the authentication credential storage and processing. Other subscription information is also stored in UDM.

The following main reference points can be identified, with a full list available in 3GPP TR 23.799.

NG1: Reference point between the UE and AMF, reflecting basically the NAS signalling between UE and AMF.

NG2: Reference point between the (R)AN and AMF, corresponding to the signalling from the 5G RAN (in some cases could also be from the LTE RAN) to the AMF.

NG3: Reference point between the (R)AN and UPF, over which user plane data is carried between the RAN and core. This has similarity to the S1-U interface with LTE, as covered earlier in this chapter.

NG4: Reference point between the SMF and UPF.

NG5: Reference point between the PCF and an Application Function (AF).

NG6: Reference point between the UPF and a Data Network (DN). As shown in Figure 2.5, the DN may be local in case of services that are carried out locally or have requirements to be carried out locally due, for example, to tight delay requirements.

NG7: Reference point between the SMF and PCF.

NG8: Reference point between UDM and AMF. This allows the fetching of user data and session data.

NG9: Reference point between two core UPFs.

NG10: Reference point between UDM and SMF.

NG11: Reference point between the AMF and SMF.

NG12: Reference point between the AMF and Authentication Server Function (AUSF).

NG13: Reference point between UDM and AUSF.

Not all reference points were fully defined, but Release 15 already included the interfaces between core and radio, which could be mapped to an LTE architecture such as NG3, which corresponds to S1_U. Naturally, NAS signalling will also be fully defined corresponding to reference point NG1, though this signalling is carried out over the radio interface via 5G-RAN to AMF.

Operation together with LTE radio/core is an essential element, especially with first-phase networks. Figure 2.6 illustrates the standalone and non-standalone architecture options. The standalone architecture is most illustrative of the case of a full 5G system with both 5G radio and 5G core operation, while Option 3 with LTE anchor and use of 5G radio only with the LTE-based legacy EPC core represents the first-phase solution. Further architecture options were completed later in Release 15, in so-called 'late drop', with the test case development for UE conformance later in the 2021/2022 time frame.

For the non-standalone case there exist further alternatives, depending how the user plane data is going to be routed. In the case of routing the data only via LTE, the user plane connection would always come via LTE, while another alternative would be to route the user plane data via 5G only. In the case of not desiring the data stream to come from both, separate user plane connections could be established to both LTE and 5G base stations, as shown in Figure 2.7.

The non-standalone operation requires the use of an X2-type interface between LTE eNodeB and 5G BTS (gNodeB, or gNB). This was

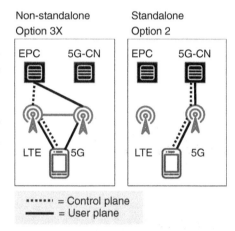

Figure 2.6 5G architecture options supported in devices 2019/2020.

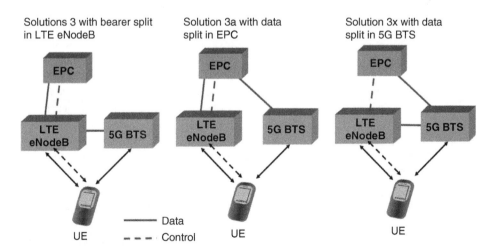

Figure 2.7 Different options for user plane data routing with dual connectivity between 5G and LTE, where LTE is the anchor to the core.

defined in 3GPP Release 15 in the same way as the X2 interface is defined today, enabling multi-vendor operation. With solutions 3 and 3x there is the split bearer option enabled, thus requiring continuous data-stream routing over the X2 interface. This needs to be taken into account in the backhaul network dimensioning as well.

The 5G gNB consists of two elements, with the Distributed Unit (DU) being at the RF site with antennas and the Central Unit (CU) at a more central location. Figure 2.8 shows the CU and DU in the RAN architecture. One CU can handle multiple DUs. The radio protocols in Release 15 defining a higher-layer functional split are distributed as shown in Figure 2.8. The DU handles the time-critical functionality as follows:

- Physical layer covering FFT/IFFT, modulation, L1 channel construction, channel coding, physical layer procedures and measurements.
- Medium Access Control (MAC) layer handling the multiplex logical channels, performing HARQ, handling CA as well as managing scheduling or new functionalities like beam management.
- Radio Link Control (RLC) layer, taking care of error correction and segmentation. When physical layer retransmission fails, RLC layer retransmissions can be used if allowed by the delay budget.

The CU takes care of less time-critical processing tasks, covering the following protocols:

- Service Data Adaptation Protocol (SDAP), handling QoS flows and their mapping on data radio bearers. SDAP enables the new QoS concept, as discussed later in this chapter.
- Packet Data Convergence Protocol (PDCP), performing header compression, security operations and guaranteeing in-order delivery without duplicates. Lower layers (RLC) provide packets to the PDCP without reordering, which enables packet duplication using multiple links (in Release 16, even up to four links).
- Radio Resource Control (RRC), taking care of system information broadcast, connection establishment, connection control, mobility and measurements.

With the interfaces between the RAN and core, communications are built using a User Datagram Protocol (UDP) with GPRS Tunnelling Protocol (GTP) for the data and Stream Control Transmission Protocol (SCTP) for the control signalling, as shown in Figure 2.9.

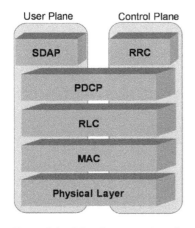

For the 3GPP based F1 interface, the same set of protocols are used as between the radio access network and 5G core. The CU functionality can be split between the control and user plane functionalities. In this case there is also an E1 interface defined in 3GPP between CU-C and CU-U parts. The E1 interface is only for control signalling, running over SCTP as well. The F1 interface is shown in Figure 2.10.

Besides the 3GPP-defined higher-layer functional split (with L1, MAC and RLC) in DU, there is also another approach defined in CPRI [1], and further addressed outside 3GPP in O-RAN (Open RAN), which covers different approaches for lower-layer functional split. The eCPRI follows the original CPRI interface, which is based on sending I/Q samples and

Figure 2.8 5G radio protocol stack.

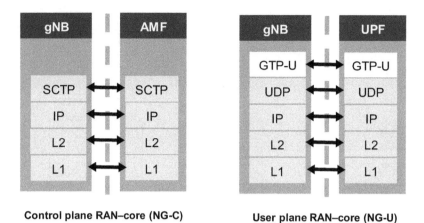

Control plane RAN–core (NG-C)   **User plane RAN–core (NG-U)**

Figure 2.9 5G protocol stack for RAN–core interface.

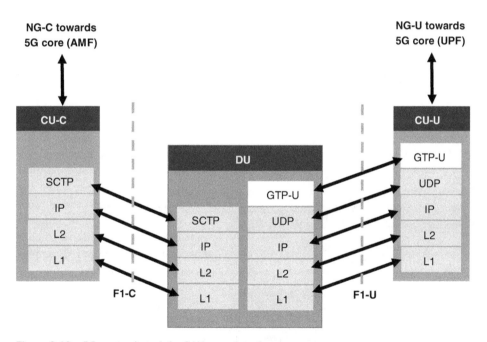

Figure 2.10 5G protocol stack for RAN–core interface.

results in very high data rates when the number of antennas increases. The latest eCPRI specification was released in August 2019 and contains multiple split alternatives. Common to all these options is that one is not transmitting I/Q samples but bits inside the physical layer. This reduces the necessary bandwidth compared to I/Q samples, while more processing happens higher up in the network. The different L1 splits shown in Figure 2.11

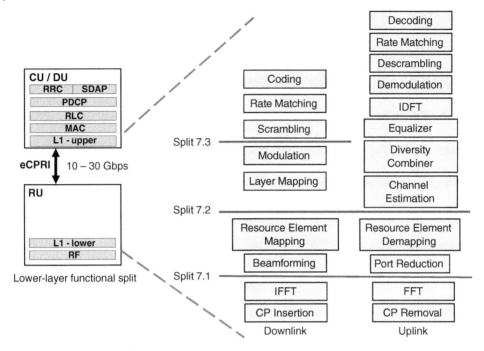

Figure 2.11 5G eCPRI different L1 split options.

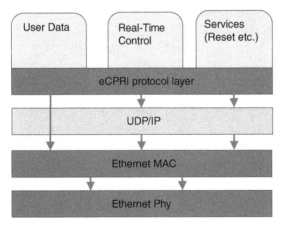

Figure 2.12 5G eCPRI protocol stack for eCPRI services.

represent different options in eCPRI, but options 7.2 and 7.3 are the most practical ones since they leave the beamforming operation to the radio side (RU). Still, with 7.1 the required transmission bandwidth is smaller than when sending pure I/Q samples with the CPRI approach.

The eCPRI itself can be carried over the Ethernet or IP layers, which thus allows both alternatives as shown in Figure 2.12. In case of Ethernet, the UDP is not used for the user place as shown in Figure 2.12. The IP alternative (IPv4 or IPv6) can then be carried over the Ethernet also, though other approaches are possible in principle. An important difference over the I/Q samples is now more tolerance to transmission path routing and delay changes, even if the requirements are rather tight; the same kind of dark fibre is no longer needed as with CPRI, but a packet-based/switched network can be used. Besides the elements in Figure 2.12, there is also additional functionality (not all defined in detail in eCPRI v 2.0) for:

- Synchronization for frame and time alignment
- Connection O&M
- Control and management information exchange

As mentioned earlier, more work on the fronthaul interface option 7.2 is being carried out on O-RAN, as well as other interfaces not covered in the 3GPP specifications. See later in this chapter for more about the ongoing development with O-RAN.

2.4 Spectrum and Coverage

The spectrum is the key asset for the operator. The available spectrum defines a lot about the network's maximum capacity and coverage. The spectrum also tends to be very expensive. This chapter discusses the expected spectrum for the early phase of 5G deployments and the characteristics of the different spectrum options. Figure 2.13 illustrates the typical spectrum usage for 5G. The mainstream spectrum below 6 GHz globally will be the mid-band spectrum at 2.5–4.9 GHz. The spectrum around 3.5 GHz is attractive for 5G because it is available globally and the amount of spectrum is high. 5G at 3.5 GHz can utilize existing base station sites. The target is to match the coverage of existing LTE1800/2100 with massive MIMO beamforming at 3.5 GHz, leading to practically full urban coverage with 5G.

5G will also need to have low bands at 600–2500 MHz in order provide deep indoor penetration and large coverage areas. Extensive coverage is important for new use cases like the IoT and critical communication. The low band could be 600 MHz, 700 MHz, 1800 MHz or 2100 MHz, or any other existing LTE band.

Figure 2.14 illustrates the outdoor coverage difference for the different frequencies compared to the 1800 MHz uplink. The calculation assumes an Okumura–Hata propagation model, downlink 3 dB better than uplink and massive MIMO gain of 6 dB compared to 2×2MIMO. This calculation shows that 5G at 3500 MHz downlink with massive MIMO can match 1800 MHz uplink coverage. The uplink at 3500 MHz does not match exactly 1800 MHz, therefore aggregation with a low band like 700 MHz would be beneficial. The calculation

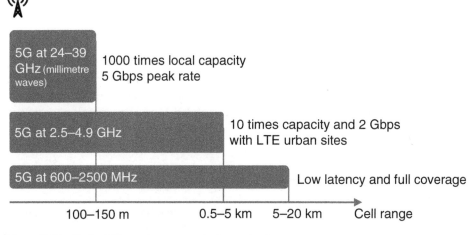

Figure 2.13 Typical 5G spectrum usage in the early phase.

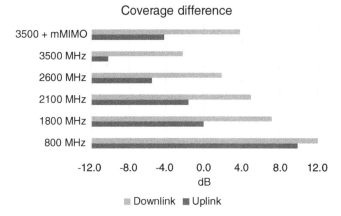

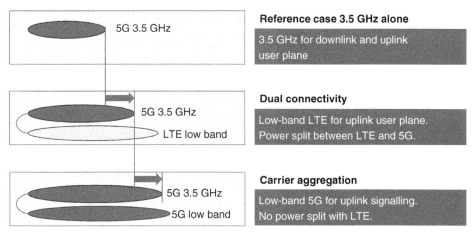

Figure 2.14 Coverage difference compared to 1800 MHz uplink.

Figure 2.15 Mid-band (3.5 GHz) 5G coverage extension using low band (LTE/5G) uplink.

shows that 3500 MHz is an attractive band for 5G and also that massive MIMO is an essential technology component in 5G.

Figure 2.15 illustrates the case where low band uplink can enhance the downlink coverage of 5G at 3.5 GHz. If uplink uses the 3.5 GHz band, it is clearly a bottleneck for coverage because the typical UE power is just 0.2–0.4 W while the base station power can be 200W or even more. The first-phase 5G is based on dual connectivity, where the uplink user plane can use low-band LTE to improve the coverage. The second-phase 5G enables carrier aggregation also, where all the uplink transmission – including signalling – can use a low band which further improves the uplink coverage compared to dual connectivity.

2.5 Beamforming

Beamforming is an attractive solution for boosting mobile network performance. Beamforming can provide higher spectral efficiency, which brings a lot more capacity to the existing base station sites. Beamforming can also improve link performance and provide an extended coverage area. Beamforming has been known in academic studies for many years, but was not supported by the first LTE specifications in 3GPP Release 8. Beamforming is included in 5G specifications already in the first 3GPP Release 15. Beamforming benefits are obtained in practice with massive MIMO antennas. The target is to make 5G radio design fully optimized for massive MIMO beamforming. The underlying principle of beamforming is illustrated in Figure 2.16. The traditional solution transmits data over the whole cell area, while beamforming sends the data with narrow beams to the users. The same resources can be reused for multiple users within a sector, interference can be minimized and the cell capacity can be increased.

Massive MIMO is the extension of traditional MIMO technology to antenna arrays having a large number of controllable transmitters. 3GPP defines massive MIMO as more than eight transmitters. Beams can be formed in a number of different ways to deliver either a fixed grid of beams or UE-specific beamforming. If the antenna has two transceivers (TRX) branches, it can send with two parallel streams to one UE. If the antenna has 4TRX, it can send four streams to one UE having four antennas, or dual stream to two UEs simultaneously with Multi-user MIMO (MU-MIMO). If the antenna has 64TRX, it can send data to multiple UEs in parallel. The number of TRXs is an important design factor in massive MIMO antennas. The more TRXs are used, the more beams can be generated which gives more capacity. But more TRXs also make the antenna size larger and costs higher. Another important antenna design factor is number of antenna elements. The number of antenna elements can be more than the number of TRXs.

Figure 2.17 illustrates an example antenna with 192 elements: 12 vertically, 8 horizontally and 2 different polarizations. The number of antenna elements defines the antenna gain and coverage. More antenna elements make the antenna size larger and increase the antenna gain. The spacing of the antenna elements depends on the frequency: the physical size of the antenna is larger at lower frequencies.

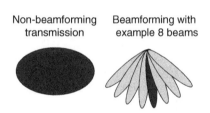

Non-beamforming transmission   Beamforming with example 8 beams

Figure 2.16 Beamforming enhances radio capacity and coverage.

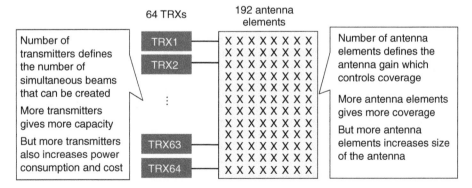

Figure 2.17 Massive MIMO principles with more transmitters and more antenna elements.

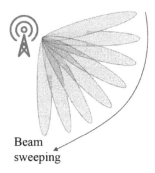

Beam
sweeping

Figure 2.18 Common
channel beamforming in 5G
with beam sweeping.

- The number of antenna elements defines antenna gain and antenna size. The size is also heavily dependent on frequency. The antenna size gets smaller at high bands.
- The number of TRXs can be the same or lower than the number of antenna elements and it defines the capacity gain.
- The number of MIMO streams can be the same or lower than the number of TRXs. The number of MIMO streams defines the peak data rate capability and is mainly dependent on the baseband processing capability.

When the number of antenna elements is larger than the number of TRXs, the additional elements are typically added as more rows. An example massive MIMO antenna could have 192 antenna elements, 64 transmitters and support up to 16 MIMO streams. There are three rows per TRX in this antenna.

There are a number of reasons why beamforming can yield more gains in 5G than in LTE:

- Beamforming is supported for 5G common channels with beam sweeping, see Figure 2.18. Beam sweeping refers to the operation where synchronization signal and broadcast channel are transmitted in different beam time domains. Common channel beamforming is not supported in LTE.
- User-specific reference signals in 5G support user-specific beamforming. LTE must use cell-specific reference signals that cannot be used for beamforming.
- No legacy device limitations in 5G, since beamforming is included in 5G from the first specifications. LTE beamforming must be based on uplink Sounding Reference Signal (SRS) measurements, since legacy devices do not support beamforming feedback.
- More transmission branches are supported in 5G. 5G initially supports feedback for 64TX, while LTE supports 4TX in Release 8, 8TX in Release 10, 16TX in Release 13 and 32TX in Release 14.

2.6 Capacity

2.6.1 Capacity per Cell

High capacity is the main solution to deliver more data to customers with the same or lower cost. Figure 2.19 illustrates a simplified view of 5G cell capacity compared to 4G cell capacity. The assumption is that 5G UE at 3.5 GHz can access 100 MHz of bandwidth while LTE1800 bandwidth (and maximum LTE single-carrier bandwidth) is 20 MHz. The spectral efficiency is improved by a factor of 2× to 5×. The cell capacity will be increased by a factor of 10× to 20×.

2.6.2 Capacity per Square Kilometre

The very high traffic density, up to 1 Tbps (1000 Gbps)/km^2, is one of the 5G targets. Traffic density can be increased by using more spectrum, more sites and higher spectral efficiency.

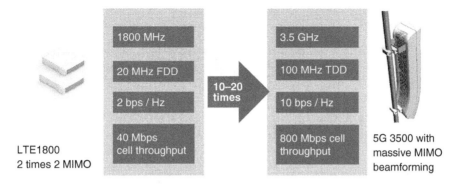

Figure 2.19 5G vs 4G capacity per cell.

5G brings all these components together for increased capacity. Let's start by estimating the highest traffic densities in the current LTE network. We exclude mass events from this calculation. We make the following assumptions for traffic estimation: 100,000 population/km^2, operator market share 40%, 5–10 Gb/sub/month usage and busy hour usage 7%. This simple calculation gives 1–2 Gbps/km^2. The typical LTE network in the busy area uses 50 MHz of spectrum and 30 base station sites/km^2. We further assume an LTE spectral efficiency of 2 bps/Hz/cell. We also take into account that the traffic is not equally distributed between the sites. The assumption is that the highest loaded sites carry 4× more traffic than the average site. These assumptions give a capacity of 2 Gbps/km^2, matching the highest traffic densities.

The capacity can be increased by fully utilizing the existing frequencies below 6 GHz for 5G. We assume that up to 200 MHz spectrum per operator is available, including some unlicensed 5 GHz band. We also assume that the site density can be increased from 30 to 50/km^2 and the spectral efficiency can be increased by a factor of 3 compared to LTE-Advanced. We also assume that the traffic distribution becomes less equal with higher site density. The result is a capacity up to 13 Gbps.

If the site density needs to be increased further, the cell range goes down to 50–100 m and it makes sense to utilize mmWs. We assume 800 MHz of spectrum and 150 sites/km^2 at 24–39 GHz. This deployment gives more than 100 Gbps/km. Spectrum above 50 GHz gives access to even more spectrum and higher site density. We assume 3000 MHz and 300/km^2, corresponding to a cell range below 50 m. This deployment case gives approximately 1 Tbps/km^2. See Table 2.2 and Figure 2.20.

Table 2.2 Traffic density assumptions.

	LTE today	5G at <6 GHz	5G at 24–39 GHz	5G at >50 GHz
Spectrum	50 MHz	100 MHz	800 MHz	3000 MHz
Sites	30/km^2	50/km^2	150/km^2	300/km^2
Sectors per site	2.5	2.5	2.5	2.5
Spectral efficiency	2.0 bps/Hz	6.0 bps/Hz	6.0 bps/Hz	6.0 bps/Hz
Traffic distribution	4	6	15	15
Traffic density	1.9 Gbps/km^2	13 Gbps/km^2	120 Gbps/km^2	900 Gbps/km^2

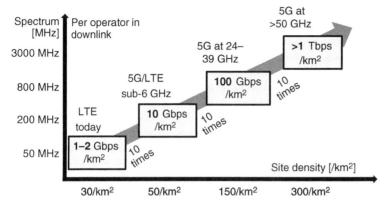

Figure 2.20 Maximum traffic density per square kilometre.

2.7 Latency and Architecture

Lower latency is an important factor in improving end-user performance, in addition to higher data rates. The target in 5G radio is to provide a sub-1 ms round-trip time. This is challenging. High-Speed Packet Access (HSPA) networks can provide 20 ms latency in the best case and the current LTE networks 10 ms. The improvement from 3G to 4G was 2×, while the target is to improve latency by 10× from 4G to 5G. The main solution for minimizing latency is shorter Transmission Time Interval (TTI). A shorter TTI makes the transmission time shorter but also makes the buffering time shorter and the processing times shorter. At the same time, the shorter processing time sets higher requirements for the receiver hardware and software. The latency components are shown in Table 2.3 and the latency evolution is illustrated in Figure 2.21. The best-case measurements with a

Table 2.3 Round-trip time components.

	HSPA	LTE	5G (3.5 GHz TDD)	5G (low-band FDD with mini-slot)
Downlink transmission	2 ms	1 ms	0.5 ms	0.14 ms
Uplink transmission	2 ms	1 ms	0.5 ms	0.14 ms
Buffering	2 ms	1 ms	0.5 ms	0.14 ms
Scheduling	1.3 ms[2]	18 ms[1]		
UE processing	8 ms	4 ms	0.8 ms	0.25 ms
BTS processing	3 ms	2 ms	0.8 ms	0.25 ms
Transport + core	2 ms (including RNC)	1 ms	0.1 ms	0.1 ms (local content)
TDD frame impact	—	—	1.2 ms	—
Total	**20 ms**	**10–28 ms**	**4.4 ms**	**1.0 ms**

1) Scheduling period + capacity request + scheduling decision + PDCCH signalling.
2) Just Shared Control Channel (SCCH).

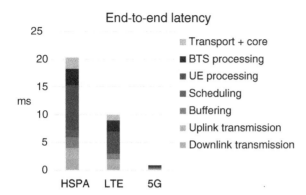

Figure 2.21 Round-trip time evolution from 3G to 5G.

Figure 2.22 Example speedtest measurements in a live network in Finland.

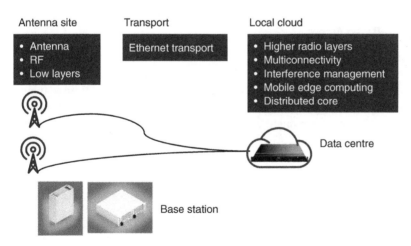

Figure 2.23 Network architecture with local cloud.

smartphone speedtest application in a live network in Helsinki are shown in Figure 2.22: 13 ms in LTE and 6 ms in 5G. The measured latencies are quite close to theoretical values. The measured 5G system uses 30 kHz subcarrier spacing and 2.5 ms TDD frame.

The future network architecture is likely to include an edge cloud, where the content can be cached for MEC or where the local breakout can be provided to the local intranet or internet. The number of local clouds needs to be more than the number of Base Station Controllers (BSCs) or Radio Network Controllers (RNCs) today. Low-latency transport is preferred from the base station site to the local cloud. The network architecture is shown in Figure 2.23.

2.8 Protocol Optimization

5G latency in the connected state is improved with mini-slot and short TTI. There are further delay components that need to be addressed in 5G radio. If LTE UE is in an idle state, there is an additional 100 ms latency caused by the establishment of RRC connection and setup of the enhanced Radio Access Bearer (eRAB). When RRC connection is available but no uplink resources are allocated, UE must send a capacity request to the base station to get a capacity allocation, causing a latency of 30 ms for the first packet. This means that the first-packet transmission in LTE typically experiences latency of 30–100 ms; a lower latency of 10–15 ms is achieved only when the uplink resources are already available. 5G addresses these latency components with a new RRC structure and by additionally defining a non-scheduled uplink transmission, also known as contention-based access.

2.8.1 Connectionless RRC

The connectionless solution in 5G refers to the case where UE can maintain the RRC connection and core network connection all the time. The UE power consumption is minimized by introducing a new 'RRC connected inactive' state. The latency will be very low because there is no need to create an RRC connection or eRAB. Current LTE networks typically release the RRC connection after 5–10 s of inactivity to minimize UE power consumption. The connectionless solution is illustrated in Figure 2.24.

2.8.2 Contention-Based Access

Contention-based access refers to the uplink transmission where UE autonomously sends a small amount of data together with a preamble without any allocation or grant from the network. The benefit of this approach is minimized signalling, which is good from a latency and UE power consumption point of view. The concept of contention-based access is shown in Figure 2.25. The preamble might be used as a reference signal or transmitted in a

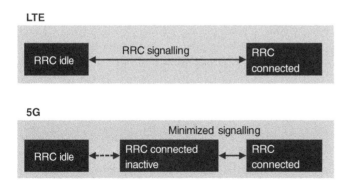

RRC = Radio Resource Control

Figure 2.24 Connectionless 5G solution with RRC connected inactive state.

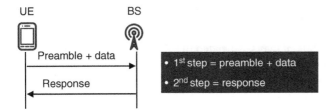

Figure 2.25 Contention-based access.

separate resource. The preamble might be associated with the UE identity or associated with a resource index for data transmission or a hopping pattern. The response might be UE specific or group-UE specific.

One option to implement contention-based access is to apply Non-Orthogonal Multiple Access (NOMA). The underlying idea here is to loosen the paradigm of orthogonal transmissions by allowing different users to concurrently share the same physical resources, in either time, frequency or space. The NOMA concept is under study in 3GPP for 5G. One could note that the WCDMA uplink is essentially based on NOMA because it is non-orthogonal. The uplink transmission is simple in WCDMA when no time alignment or exact scheduling is required, but uplink interference has turned out to be a major issue in WCDMA in mass events with lots of uplink traffic. It is clear that the NOMA concept requires an efficient uplink interference cancellation in the base station.

2.8.3 Pipelining

Another solution for minimizing device power consumption is pipelining, which allows the UE to utilize micro-sleep power saving during data symbols if there is no data transmission to the UE. Pipelining also supports low-latency transmission because the data can be decoded symbol-to-symbol continuously. The LTE UE must receive the whole sub-frame before decoding can be started. For larger UE velocities, an additional Demodulation Reference Signal (DMRS) has to be added. See Figure 2.26.

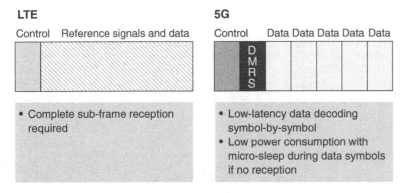

Figure 2.26 5G pipelining receiver solution.

2.9 Network Slicing and QoS

A fundamental rethink of the mobile network is needed in 5G to support very diverse and extreme requirements for latency, throughput, capacity and availability. The network architecture must shift from the current 'network of entities' architecture to a 'network of capabilities' architecture. The network models must shift from the current 'network for connectivity' model to a 'network for services' model. Network slicing offers an effective way to meet the requirements of all use cases in a common network infrastructure. The concept of network slicing is illustrated in Figure 2.27.

The 5G network needs to have effective means for network slicing. LTE supports Quality of Service (QoS) differentiation but we need dynamic application-based Quality of Experience (QoE) in 5G. This approach is not achievable in LTE, where the same QoS is applied for all traffic within a bearer. The difference between LTE QoS and 5G QoS is illustrated in Figure 2.28 and the difference between bearer-based QoS and dynamic QoE in Figure 2.29. The bearer-based solution is fine for operator-provided services, where the packet filters are easy to define and application sessions are long lived. The 5G QoE architecture must detect and differentiate short-lived sub-service flows. The control plane signalling of packet filter attributes and related policies is not necessary when both radio

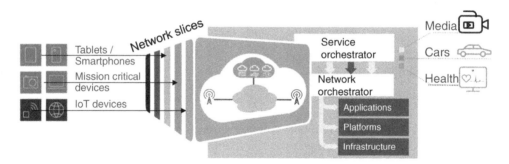

Figure 2.27 Network slicing concept.

LTE Baseline for QoS	5G Baseline for QoE
In EPC/LTE, the QoS enforcement is performed at the eNodeB for uplink and Policy and Charging Enforcement (PCEF) for downlink. Policy decisions are taken in the core network.	In 5G, a framework for **end-to-end QoS/QoE** is built into the baseline architecture.
QoS differentiation is achieved by enforcing QoS targets such as the delay budget, guaranteed bit rate and relative throughput ratio among bearers.	Both radio and core network elements have capabilities for real-time application awareness, QoE performance awareness and intelligence for dynamic policy modifications.
The radio and core network enforce QoS independently, uplink and downlink QoS are not coordinated.	Both radio and core elements are able to track the unidirectional performance of application flows and take enforcement actions in both directions.

Figure 2.28 From LTE quality of service (QoS) to 5G quality of experience (QoE).

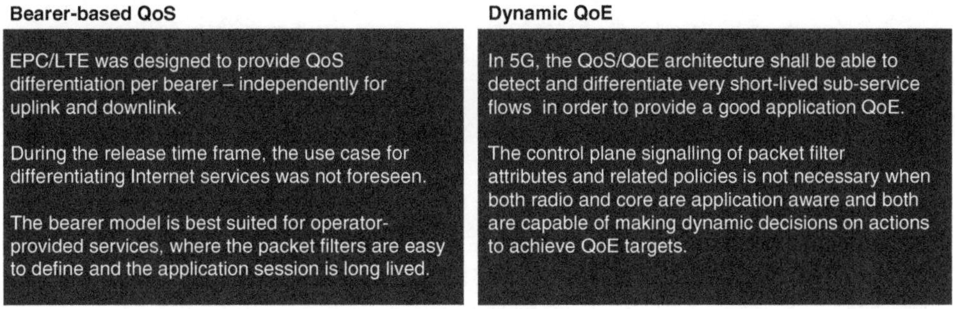

Bearer-based QoS

EPC/LTE was designed to provide QoS differentiation per bearer – independently for uplink and downlink.

During the release time frame, the use case for differentiating Internet services was not foreseen.

The bearer model is best suited for operator-provided services, where the packet filters are easy to define and the application session is long lived.

Dynamic QoE

In 5G, the QoS/QoE architecture shall be able to detect and differentiate very short-lived sub-service flows in order to provide a good application QoE.

The control plane signalling of packet filter attributes and related policies is not necessary when both radio and core are application aware and both are capable of making dynamic decisions on actions to achieve QoE targets.

Figure 2.29   From bearer-based QoS to dynamic QoE.

and core are application aware and both are capable of making dynamic decisions on actions to achieve the QoE targets.

The main features of the 5G QoE architecture are as follows.

- Cognitive functions in radio access network and core network: advanced real-time analytics for anomaly/degradation detection, context evaluation.
- Dynamic and adaptive, policy-based operation: high-level policies applied by considering the context – like location, traffic mix, user behaviour and network status.
- Real-time application detection, context-based QoE target definition and QoE enforcement: operation and enforcement actions adapted to the specifics of the application sessions.
- Harmonized, end-to-end QoE including transport and virtual resources.
- Insight-driven radio scheduling: enhanced insight is shared with the radio functions allowing improved efficiency and consistency in the system operation.
- Capability to operate according to traditional QoS and network-level targets as well.
- The QoS architecture has developed further from LTE. There is no longer a user/bearer-specific tunnel established between the core and radio to handle a specific quality of service bearer. Instead, the encapsulation header on the NG3 interface (user plane interface between RAN and core, as covered earlier) is used to carry the marking for the QoS to be used on the downlink packets. The UE will then apply the same quality of uplink packets corresponding to service with a given quality in the downlink direction. This is referred to as reflective quality. The overall quality framework is illustrated in Figure 2.30 The approach chosen in 5G allows fast reaction to application needs based on the intelligent core, and dynamical reaction to changing needs for best possible user experience.

The UE will be provided with a default QoS rule at PDU session establishment via NAS signalling. Also, 5G RAN is provided with QoS profiles of the QoS rules from NG-core. The QoS parameters include the following:

- UL and DL Maximum Flow Bit Rate.
- UL and DL Guaranteed Flow Bit Rate.
- Priority level.

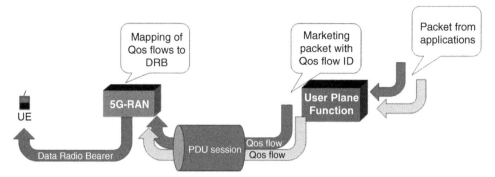

Figure 2.30 QoS architecture in 5G.

2.10 Integrated Access and Backhaul

As part of 3GPP Release 16, support for Integrated Access and Backhaul (IAB) was introduced. The basic idea with IAB is to address situations where providing traditional backhaul is difficult, thus the possibility to use the transmission band for backhaul transmission as well is considered. This is especially targeted at mmW operation, such as the 28 GHz band, where generally operators have a great deal of spectrum available. With IAB operation the data is relayed from the donor gNB to the IAB gNB over the 3GPP radio interface. As shown in Figure 2.31, there can be more than a single hop. Using many hops adds, of course, a delay for the connection and reduces the capacity available for end users, especially at the first connection point.

The IAB architecture has introduced Mobile Termination (MT) on the IAB node side, as shown in Figure 2.32. MT basically covers the functionality that is not part of a regular gNB, namely capabilities such as cell search, capability to create a connection starting from the RACH procedure towards a donor gNB, as well as providing feedback on channel conditions or for MIMO and beamforming operations.

The use of IAB allows us to avoid having fibre to every radio site, but comes with other implications:

- Increased latency, as each hop adds more latency for the connection
- Additional coordination needs if the same band in the same location is used by another operator for regular or IAB use
- Part of the band capacity is occupied for backhaul needs

From the UE perspective, there is no visibility on whether a gNB is a regular gNB or behind an IAB connection. Increased latency and reduced data rates become visible, but the UE can't tell if those are due to other reasons or due to the use of IAB backhaul. Practical deployment may have to consider aspects of coordination between operators if the same band is being used by IAB nodes in the same locations by different operators, to avoid the known issue of TDD uplink/downlink interference.

There is work ongoing in Release 17 to further evolve the work with IAB, including more consideration of duplexing aspects such as simultaneous operation of the IAB node's child

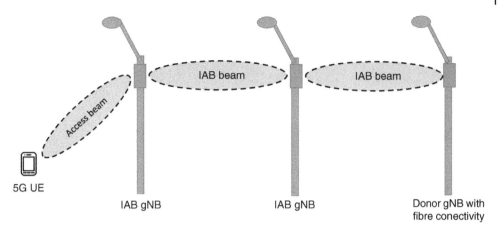

Figure 2.31 IAB operation principle.

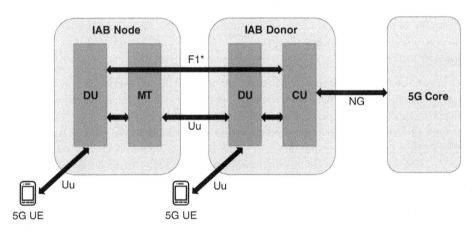

Figure 2.32 IAB architecture.

and parent links, or enabling dual connectivity for topology redundancy, as well as enhancements to reduce service interruption due to IAB-node migration and backhaul Radio Link Failure (RLF) recovery.

2.11 Ultra Reliable and Low Latency

In Release 15, the first elements for Ultra Reliable and Low Latency (URLLC) were included in the 5G specifications. This allows us to deliver service with higher reliability than what is needed for the mobile broadband use cases. Release 15 was aiming for 10E-5 reliability (e.g. suited to entertainment use cases), while Release 16/17 was further targeting a more reliable (10E-6) service delivery for transportation, factory automation and energy distribution control requirements support. The elements already in Release 15 include:

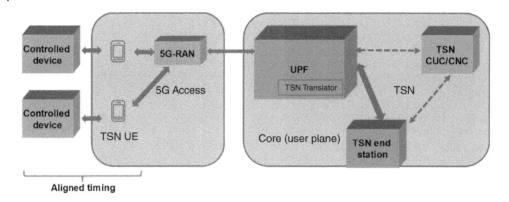

Figure 2.33 TSN operation over 5G network.

- High-reliability Modulation and Coding Set (MCS) tables and corresponding Channel Quality Information (CQI) feedback definitions, as well as low-latency operation with mini-slot. Further, there is also support for PDCP-level data duplication and logical channel priorization.

In Release 16 the work continued for the enhancement of URLLC as well as for Industrial IoT operation. This work focused on link improvements for better reliability, like PUSCH repetition and addressing intra-UE priorization, but also addressing higher-layer improvements with Ethernet header compression as well as extending the PDCP duplication up to four links. Overall, this should enable use of Time-Sensitive Networking (TSN) where the transmission timing is deterministic, enabling support of IEEE 802.1-type Ethernet operation with 5G. This is valid for use with industrial automation, with the importance to have, for example, robots acting on the same object with synchronized timing. An example of TSN is shown in Figure 2.33.

Release 17 expanded this beyond TSN with Time-Sensitive Communications (TSC), allowing support for a wide range of factory automation solutions on top of 5G with enhanced time-synchronous operation support. The solutions allow us to provide better timing synchronization, including addressing the propagation delay compensation, thus making 5G RAN a better platform for the different factory automation solutions to run on. The work for Release 17 also covered improvements in URLLC operation, with an unlicensed spectrum with Frame-Based Equipment (FBE) mode of unlicensed band operation to allow for the UE-initiated Channel Occupancy Time (COT) for lower latency.

2.12 Open RAN

As indicated previously, more work on the fronthaul interface between the radio site and the remote site is being carried out in the O-RAN alliance [2]. While 3GPP basically introduced the split between the CU and the DU, the O-RAN specification introduced a further split between DU and RU functionalities. The functional split chosen by O-RAN is option 7.2 defined in the eCPRI spefications discussed earlier. O-RAN defines more detailed open-interface specifications between RU and DU, as not all aspects of the interface necessary for multi-vendor operation were covered in the eCPRI specifications. In 2022 O-RAN started further work on DU/RU functional split evolution for improved uplink performance. Additionally, there are some 10 other interfaces being addressed by the O-RAN work.

In the O-RAN environment, fronthaul (FH) is used to refer to the connection between RU and DU, while midhaul is used for the F1 connection between DU and CU, as shown in Figure 2.34. In this case, the functional split between CU and DU is as defined in 3GPP. Besides 3GPP F1 (also called 3GPP higher-layer split), O-RAN-based deployment or classical-type deployment all have BTS functionality on the radio side, as is typical with LTE deployment, unless approaches like baseband hotels are used. The 5G deployments were started before the O-RAN specifications were available, thus O-RAN-based FH (or other interfaces) are expected to be introduced in the later phase of 5G network deployment. Which of the fronthaul solutions are used has impacts for backhaul requirements, including how far the next unit can be located from the actual radio site. These aspects are addressed in later chapters of this book.

Besides the RU and fronthaul interface shown in Figure 2.34, O-RAN introduces other possible elements such as the RAN Intelligence Controller (RIC), which is aimed at optimizing network performance. The RIC, as shown in Figure 2.35, is a control plane element and connects to the CU with a control plane-only interface. User data itself doesn't go via the RIC. For the control plane interface to the RIC, the name E2 has been used in O-RAN work [2]. The RIC is expected to host different (near real-time) applications that would help the CU operation in different areas of radio resource management. The RIC itself would then further connect to the orchestration and automation functionality of the network for non-real-time operation/automation. The interfaces based on O-RAN specifications are complementary to the 3GPP-defined interfaces.

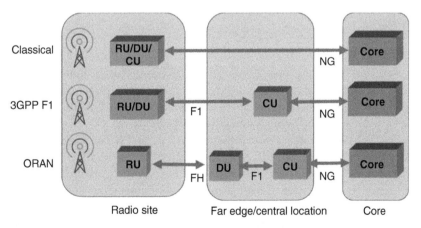

Figure 2.34 Different fronthaul/midhaul deployment options.

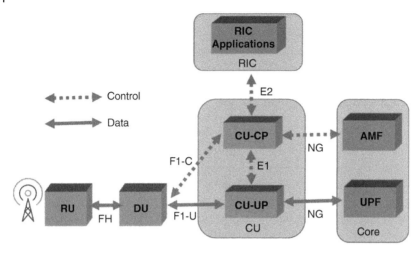

Figure 2.35 RIC in O-RAN architecture.

2.13   3GPP Evolution in Release 16/17

The 3GPP standards delivered the first full set of specifications (including UE capabilities) for the NSA mode in December 2018, followed by standalone (SA) 5G specifications in 2019. After Release 15, the effort for the rest of 2019 focused on Release 16 work, which was mostly finalized at the end of 2020. The content (at feature level) has also been agreed for Release 17, with the work started during 2020 being finalized in June 2022 (implementable by September 2022), as shown in Figure 2.36.

The key new elements in Release 16 are shown in Figure 2.37, including Industrial IoT as well as 5G support for unlicensed band operation (5 GHz unlicensed band) as well as features like 5G V2X or UE power saving or MIMO enhancements. The first versions of IAB were also introduced in Release 16. More details of the different Release 15/16 features can be found in Ref. [3].

Release 17 continued to expand the ecosystem, while further improving mobile broadband operation in many areas. The key elements in Release 17 include Enhanced Industrial IoT operation, as well as further enhanced MIMO and further enhanced positioning. Some of the new elements include RedCap (Reduced Capability), which is developing a new UE category for applications not requiring full 5G capability, such as surveillance cameras, industrial sensors or modules for home automation, where data rates of 10–100 Mbps would be sufficient. IAB enhancements were also included, as well as developing further usability of sidelinks for both public safety operations and solutions like Vehicle to Pedestrian (V2P) use cases. A new segment is also the use of NR with satellite networks, using the 3GPP term Non-Terrestrial Networks (NTN), accommodating High-Altitude Platforms (HAPS). The key Release 17 elements are shown in Figure 2.38.

There will be further elements introduced in Release 18, after the finalization of Release 17 (in mid-2022). There was some delay in the original Release 17 schedule, as during 2020 and 2021 the planned 3GPP meetings had to be handled electronically due to Covid-19

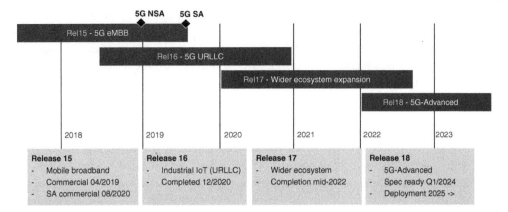

Figure 2.36 3GPP standards schedule.

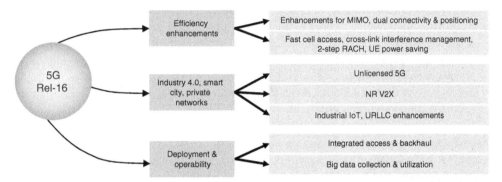

Figure 2.37 3GPP Release 16 key content.

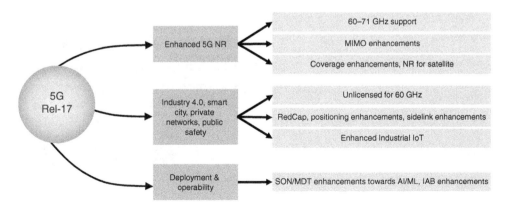

Figure 2.38 3GPP Release 17 key content.

restrictions and thus the start of Release 17 work was postponed, with a focus on the Release 16 finalization in 2020 E-meetings replacing regular 3GPP events. The Release 17 content itself remains unchanged but the schedule was delayed by 9 months, with finalization in mid-2022. 3GPP concluded on Release 18 content at the end of 2021. The latest information about the 3GPP schedule can be found in Ref. [4].

2.14 5G-Advanced

3GPP Release 18 is labelled '5G-Advanced'. This new name can be justified because Release 18 sits in the schedule between the first version of 5G and the expected first version of 6G specifications. The new name can also be justified because Release 18 capabilities extend beyond the original 5G vision. The main areas of 3GPP evolution are shown in Figure 2.39.

- Radio boosters: radio capability improvements in terms of data rates, coverage and capacity continue in Release 18. The focus moves mostly to the uplink direction, as the expectation is that future services will require higher uplink capabilities.
- Automation and energy saving: Artificial Intelligence (AI) and Machine Learning (ML) can be utilized for network optimization and later even for air-interface parameters. Energy saving is another important target to be considered in 3GPP feature evolution.
- New verticals: the target is to optimize 5G radio for new connectivity cases including car-to-car, satellites, high-altitude platforms, railway control communications replacing GSM-R, public safety and drones.
- IoT optimization: RedCap UEs will be optimized for IoT connectivity by having lower cost and lower power consumption compared to full-blown 5G modems. 5G IoT will provide substantially higher capabilities compared to LTE-based IoT solutions.
- Accurate position and time: the 5G network will be able to provide users with information about their exact location with 10 cm accuracy and exact timing information. Traditionally, satellite-based solutions are needed for this information.

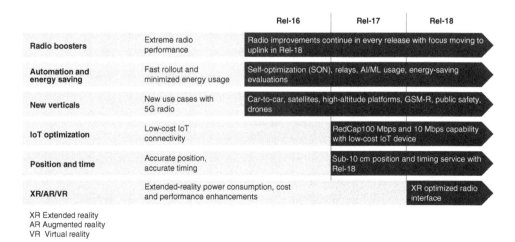

Figure 2.39 Main contents of 5G-Advanced.

- Extended Reality (XR): 5G-Advanced will bring a number of improvements for XR connectivity, including XR device power savings, latency, mobility, application awareness in the radio, edge cloud usage and capacity.

The 3GPP target is to complete Release 18 during 1H/2024, which would lead to the first commercial deployments starting from 2025 onwards.

References

1 http://www.cpri.info/spec.html (accessed 4 July 2022).
2 www.o-ran.org (accessed 4 July 2022).
3 Holma, H., Toskala, A., and Nakamura, T. (eds.). *5G Technology: 3GPP New Radio*. Wiley: New York.
4 www.3GPP.org (accessed 4 July 2022).

3

5G RAN Architecture and Connectivity – A Techno-economic Review

Andy Sutton

3.1 Introduction

Mobile network operators have many inputs to consider when deciding how to implement 5G in their networks. The use of mobile backhaul is already extensive in support of existing radio access technologies, while some operators have also implemented fronthaul for LTE. The use of fronthaul enables a C-RAN architecture which offers some economic advantages over a fully distributed (or aggregated) RAN (D-RAN) in certain markets. The economic benefits are not realized in all markets; they depend on the availability of dark fibre and flexibility of contracts with site providers. The technical performance benefits of an LTE C-RAN are realized through the implementation of real-time coordination between multiple geographically dispersed radio transceivers controlled by a common centralized baseband unit. Benefits include improved cell edge performance and reduced interference. This chapter aims to review the techno-economic aspects of 5G RAN backhaul, midhaul and fronthaul, and explore different deployment scenarios from the perspective of a mobile network operator. The chapter will, in the main, focus on 3GPP FR1, however Section 3.11 introduces FR2 in the context of small cells.

3.2 Multi-RAT Backhaul

Existing mobile backhaul networks often support GSM, UMTS and LTE backhaul and, over the last decade, have migrated from TDM-based transmission systems to Carrier Ethernet with an IP transport network layer. GSM Abis was defined as a TDM interface, however all vendors now implement the Abis interface with IP; this has never been standardized by 3GPP so is purely proprietary. UMTS was initially implemented as an ATM transport network layer. However, this was later defined as IP in 3GPP Release 5. Most operators have now migrated to IP and changed their underlying transmission network from TDM to Carrier Ethernet. LTE was specified from the outset, within 3GPP Release 8, to have an IP transport network layer. The backhaul strategy and architecture developed for LTE will often be the starting point for an operator to review what needs to change to

5G Backhaul and Fronthaul, First Edition. Edited by Esa Markus Metsälä and Juha T. T. Salmelin.
© 2023 John Wiley & Sons Ltd. Published 2023 by John Wiley & Sons Ltd.

add 5G to their network. Figure 3.1 illustrates a typical pre-5G mobile backhaul network which supports a GSM Abis interface, UMTS Iub interface and LTE S1 and X2 interfaces.

Figure 3.1 includes a RAN aggregation site which accommodates the GSM BSC, UMTS RNC and IP Security Gateway, which is often used by operators to secure their LTE backhaul and, if using GSM IP Abis, the security gateway will likely be used to secure the Abis interface too. The RAN aggregation site will provide onwards connectivity to the respective core network elements for the various radio access technologies, however this is beyond the scope of this chapter and therefore is not illustrated in Figure 3.1. The focus of this chapter is on the network elements within the RAN connectivity site and transport network to the left of the RAN aggregation site; this is the backhaul network. The network controllers and security gateway are connected to a Mobile Aggregation Site Gateway (MASG) or Provider Edge (PE) router (IP/MPLS network edge router), which aggregates the backhaul circuits and provides a range of layer 2 and layer 3 services towards the radio access network. During the evolution from TDM to Carrier Ethernet backhaul it was very common for the MASG/PE to support pseudo-wires, however nowadays it is more common for the backhaul network to be layer 3 based, typically utilizing IP VPN. To build a suitable level of resilience into the backhaul network, it is recommended to provide resilience through 1+1 operation of MASG/PE and IPsec GW platforms. Depending on the actual implementation of the network between the cell site – fibre hub and RAN aggregation site, it is recommended that resilient transmission be provided from a suitable intermediate node such that a single fibre or equipment failure doesn't cause a wide-area outage.

In Figure 3.1 the transmission path between the RAN aggregation site and the first cell site (cell site – fibre hub) is an optical fibre connection via the Optical Line Terminating Equipment (OLTE). Given the multi-RAT backhaul requirements illustrated, along with those of the subtended microwave radio-connected cell site, it is likely to be a 1 Gbps Ethernet (1GE) circuit. The different backhaul interfaces will be dimensioned according to demand within the shared 1GE backhaul circuit. The cell site – fibre hub is acting as a local point of aggregation for its co-located base stations and those connected via the subtended microwave radio link (MW), and therefore needs to be dimensioned accordingly. Figure 3.1 illustrates a simple two-site topology, however many other topologies will be found in operational networks, from linear chains to stars and tree and branch topologies. LTE has

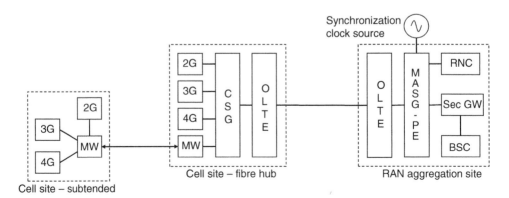

Figure 3.1 Pre-5G mobile backhaul architecture.

resulted in a greater amount of optical fibre backhaul, even in networks which have traditionally favoured microwave radio backhaul; this is driven by the need for higher capacity within the access network.

The RAN aggregation site in Figure 3.1 provides the network synchronization feed, illustrated as a clock connection to the top of the MASG/PE. For many pre-5G networks the synchronization requirement was to support the frequency of radio base station transmission and therefore delivering a frequency reference to discipline the local oscillator(s) was adequate. The FDD radio interface requires a frequency stability of 50 ppb so, to allow some margin for the base station, the backhaul network typically delivers a stability of 16 ppb. This frequency synchronization is typically traceable to an ITU-T G.811 clock source and could be delivered via the HDB3 line code in an E1 circuit or, since the backhaul migration to Carrier Ethernet [1], via Synchronous Ethernet (SyncE) [2, 3].

3.3 C-RAN and LTE Fronthaul

LTE C-RAN has been implemented in some markets. In these networks it is common to have mixed backhaul and fronthaul requirements on the access network connectivity solution, due to legacy radio access technologies (GSM and UMTS) requiring backhaul while LTE requires fronthaul for a C-RAN implementation. The most common solutions used for such hybrid backhaul/fronthaul requirements are either several pairs of fibre, one being used to connect the Cell Site Gateway (CSG) and others being used to connect to the individual LTE remote radio units, or a DWDM system in which different wavelengths are used to transport the backhaul and fronthaul services in parallel.

LTE C-RAN implements a split base-station architecture by separating the RF functionality from the physical layer and higher-layer protocols of MAC, RLC and PDCP. This is referred to as an option 8 split within 3GPP. This option 8 split interface has been defined by an industry group of vendors, known as the Common Public Radio Interface (CPRI). The vendors involved with the initial CPRI specification were Ericsson, Huawei, NEC and Nokia [4]. An option 8 split places significant demands on the transport network in terms of required data rate and low latency. As an example, a 20 MHz 2×2 MIMO LTE carrier requires 2.4576 Gbps of fronthaul capacity, an order of magnitude more than the backhaul capacity required for such a carrier, given that the typical maximum user throughput is approximately 195 Mbps (assuming 256 QAM on downlink). The latency constraints are considerable based on this split; the figures vary slightly between vendors, however a Round-Trip Time (RTT) latency of 250 μs is typical. This maximum latency is determined by the timing requirement of the Hybrid Automatic Retransmit reQuest (HARQ) protocol used as a retransmission mechanism between the UE and eNB in an LTE network. In practical systems, the latency constraint limits the maximum one-way fibre cable distance to ~20 km, this is implementation dependent.

Figure 3.2 illustrates a combined backhaul/fronthaul service being delivered from a RAN aggregation site to a cell site. The GSM and UMTS services are being carried as traditional backhaul, via the CSG, while the LTE is being provided with a fronthaul service to support three remote radio units (also known as remote radio heads). In this example, the CPRI channels are being carried over optical wavelengths via an access DWDM system.

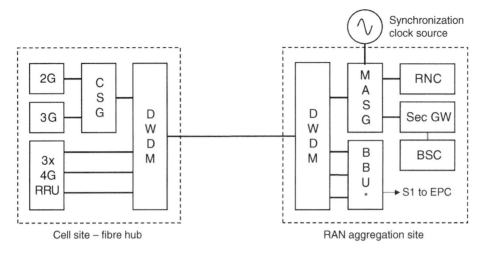

Figure 3.2 Backhaul and LTE C-RAN fronthaul.

It should be noted that the C-RAN architecture with CPRI fronthaul can be used for GSM and UMTS too, however this is less common than LTE. A wide range of transmission solutions for LTE CPRI were examined by NGMN, including microwave and millimetre-wave radio systems, in their white paper [5].

3.4 5G RAN Architecture

The 5G RAN architecture adds new inputs to the mobile backhaul/fronthaul decision-making process, while the need to interwork with LTE, in the EN-DC mode of operation (5G NSA [Non-Standalone Architecture]), adds yet more considerations. It is essential to understand how 5G changes the RAN architecture, what new interfaces are defined and how these map to available transport network technologies to enable the mobile operators to conduct a detailed techno-economic analysis to inform future strategy and target architecture.

A 5G radio base station, known as a gNB, can be decomposed into three functional entities: Radio Unit (RU), Distributed Unit (DU) and Central Unit (CU). This functional decomposition is illustrated in Figure 3.3.

The gNB consists of a radio unit which could be integrated within the antenna module, in the case of massive MIMO, or operate as a standalone radio unit, in the same manner as an LTE RRU. The baseband module is decomposed into the two remaining functions; the DU and CU. These names are a little misleading, as both could be distributed or both could be centralized or, as indicated by the names, the DU could be distributed while the CU is centralized. The advantages and disadvantages of the various options will be explored throughout the remainder of this chapter.

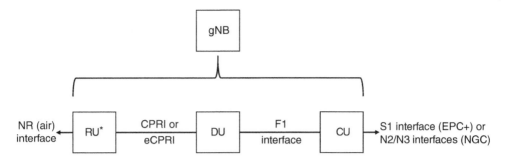

* RU could be integrated within AAU (mMIMO) or standalone RU (RRU/RRH)
with coaxial connections to passive antenna (typically 8T8R)

Figure 3.3 5G RAN functional decomposition.

The functionality within the radio unit depends on whether the interface to the DU is CPRI, as discussed earlier in relation to LTE C-RAN, or eCPRI. eCPRI is an evolved form of CPRI, which was specified by the same four vendors – Ericsson, Huawei, NEC and Nokia – in 2017 as eCPRI version 1 [6] and further updated in 2019 as eCPRI version 2 [7]. The driver for eCPRI was to reduce the fronthaul capacity requirements between RU and DU by moving part of the physical layer processing, residing in the DU, to the RU. This is referred to as the lower physical layer and results in an option 7 split which reduces the overall data rate required on the interface while also being supported over a standard Ethernet circuit. eCPRI introduces a new EtherType (AEFE) as an identifier within the Ethernet frame. eCPRI also introduces a variable bit-rate flow which aligns with the utilization of the radio interface; option 8 CPRI is a constant bit-rate signal. It is important to note that not all 5G NR RUs will support eCPRI; early deployments have eCPRI interfaces for massive MIMO antenna systems however any external RU-based systems with lower-order MIMO (8T8R or less) will typically utilize CPRI. As LTE carriers are refarmed to NR, possibly via Dynamic Spectrum Sharing (DSS), the fronthaul interface will remain as CPRI, therefore a mixed mode of fronthaul transmission will be required.

From a mobile network operator perspective, the introduction of eCPRI is a beneficial optimization of the fronthaul segment. The reduced capacity requirements relative to CPRI translate to lower-cost optical pluggable transceivers; whether the RU and DU are co-located at the cell site or whether a C-RAN is implemented. These considerations will factor into a wider economic analysis later in this chapter.

The DU contains the full physical layer in the case of an option 8 interface, however in most 5G deployment scenarios, certainly those beyond phase 1 hardware, it will likely be connected to an option 7 interface and therefore contain the higher physical layer along with lower and higher MAC and lower and higher RLC protocols. It should be noted that there are several split points being discussed for an option 7 interface; at the time of writing this is still an open topic within standards and industry forums, however the principles discussed in this chapter apply to all option 7 splits.

In an EN-DC-configured 5G network the CU contains the PDCP layer for the user and control plane along with the RRC signalling layer. Deployed with a 5G core network,

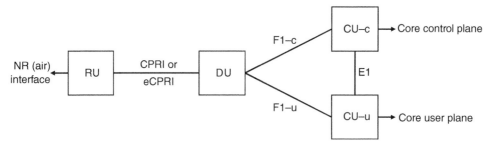

Figure 3.4   5G RAN functional decomposition with split CU.

known as Next Generation Core (NGC), the CU will also include the new SDAP protocol which sits above PDCP on the user plane. The CU is separated from the DU by the 3GPP standardized F1 interface [8], this is an option 2 split. In early 5G implementations it is common for the CU functionality to be integrated with the DU in a 5G baseband unit. The F1 interface is a non-real-time interface and therefore has no exacting latency requirements, unlike the option 7 interface which was discussed earlier. The capacity requirements of the F1 interface are similar to traditional S1 (or N2/N3) backhaul; the F1 interface requirement will be slightly higher, however only by ~5%.

The CU can be further decomposed to CU-c and CU-u functionality; this control and user plane separation builds on established architectural principles within 3GPP. This started initially in the core network with Release 4 control and user plane separation within the circuit-switched core network, and this specification resulted in the creation of a media gateway and MSC server. Similar principles have since been applied to the packet-switched core. The CU-c and CU-u functions are separated by a 3GPP-defined interface, designated as E1 (Figure 3.4).

While early implementations of 5G RAN will be either D-RAN or in certain markets C-RAN, it is likely that the DU and CU will be combined in a single baseband unit. Most vendors have gone to market with this as the only option from day 1. The DU–CU split implementation is on roadmaps and will become more common from 2022 onwards; such a split architecture is standard with an O-RAN solution, albeit practical implementation would allow DU and CU to be co-located if required.

3.5   5G D-RAN Backhaul Architecture and Dimensioning

Adding 5G to an existing multi-RAT cell site will have a significant impact. 5G rollout is proving to be a lot more than just a technical engineering challenge, it also involves lots of civil engineering to enable sites and structures to support NR equipment – radio units plus larger passive antennas and/or massive MIMO active antennas. The commercial review of implementing 5G site upgrades must include all costs, from network design and planning to civil engineering and power upgrades to NR equipment and associated installation, commissioning and integration activities.

The amount of NR spectrum an operator has will determine the overall backhaul capacity requirements, and this will determine the speed of Carrier Ethernet circuit which needs to be deployed. The NR spectrum requirements must be added to those of the existing multi-RAT site to determine the overall requirements. Given the EN-DC mode of operation, an operator may choose to deploy additional LTE spectrum as part of their 5G upgrade, to further boost peak data rates and overall cell-site capacity. There are several approaches to calculating mobile backhaul capacity requirements and care must be taken to ensure alignment with business and commercial strategy; this should include an analysis of forecasted traffic levels on a site-by-site basis for the coming years. Taking a demand-based view of capacity planning will result in specific requirements per site, and this can be very labour intensive and therefore expensive. That said, it may be valid in some networks where minimizing investment in a given year is important or in certain geographical areas where a more ambitious approach to backhaul capacity dimensioning simply can't be justified due to low traffic volumes.

The most common approach in dense urban, urban and suburban areas is to consider the deployed radio resources and calculate the peak and average throughput based on achievable spectrum efficiency. The peak spectrum efficiency will likely be close to the theoretical limits as calculated in the 3GPP standards; the average will be significantly lower as this will be constrained by interference from adjacent sites. Reducing intersite interference is a consideration for the deployment of a C-RAN architecture, be this a centralized BBU in the case of legacy RATs or a centralized DU and CU in NR. The centralization of the DU offers maximum capacity benefits through real-time coordinated scheduling, however this comes with the challenges of high-capacity and low-latency fronthaul as discussed earlier. Once the peak and average spectrum efficiency are calculated on a per RAT basis, these can be summed and associated backhaul overheads added.

Backhaul overheads will depend on the status of an operator's 5G network; whether it is non-standalone or standalone will determine whether backhaul is S1-c and S1-u or N2 and N3, likewise whether an X2 or Xn interface is in use. Overheads vary between control and user plane too, however it's the user plane that poses a specific challenge due to the large volume of traffic and range of IP packet size payloads. Taking an example of a 5G EN-DC network, the control and user planes connect between the base stations (eNB/gNB) and EPC+ and contain the protocols detailed in Table 3.1.

Table 3.1 illustrates a typical mobile backhaul protocol stack for 5G EN-DC operation; this doesn't change significantly with a 5G core network. The S1-AP in the control plane is replaced with NG-AP, while in 5G core network terminology the user data maps into a Protocol Data Unit (PDU) which is the layer 4 (transport) and layer 3 (IP) from Table 3.1 S1-u; this PDU then maps to GTP-u. The remainder of the stack is the same for the N3 interface. The use of IP Security is optional; many operators do implement this for 5G backhaul, with both features of Encapsulating Security Payload and Authentication Headers implemented. IPsec tunnel mode is most common. This adds an IP address to the packet which addresses the IPsec endpoint, therefore providing a secure end-to-end tunnel. The use of IPsec has many advantages, however it does cause some transport network optimization challenges – more on this later in the chapter. The protocol stack above assumes a Carrier Ethernet access circuit; many such access circuits map to layer 3 IP VPN services once they reach an aggregation node, and at this point the lower part of the protocol stack would change.

Table 3.1 Example of backhaul overheads of an S1 interface for 5G EN-DC mode of operation.

S1-c interface (from eNB in EN-DC mode)	S1-u interface (from gNB)
S1-AP	User data
SCTP	TCP/UDP/QUIC
IP TNL (v4 or v6)	IP (v4 or v6)
IPsec ESP/AH*	GTP-u
Outer IP (v4 or v6)*	IP TNL (v4 or v6)
VLAN	IPsec ESP/AH*
Ethernet	Outer IP (v4 or v6)*
Underlying network	VLAN
	Ethernet
*IP Security is optional	Underlying network

With an understanding of the protocols and an appreciation of the varying packet sizes which can occur, the backhaul overhead for the protocol stack above can be calculated to be 15–20% of the user data payload. Once this is understood, it can be added to the peak and average spectral efficiencies and resulting data rates for the 5G service. Three options now exist: assuming a three-cell sector site, the three peaks could be summed; the three averages could be summed; or a single peak could be added to two averages (this latter approach is known as the single peak–all average model). Summing all three peaks would result in over-dimensioning and therefore more total spend than is necessary. Summing the three averages would constrain the peak data rates when users are in favourable radio conditions, therefore impacting the perception of the network. The single peak–all average approach offers a pragmatic solution which takes the peak data rate from one cell sector and adds this to the average data rate of the two remaining sectors. This ensures a user in good radio conditions can achieve impressive peak rates while implementing a level of traffic engineering to strike a balance in the techno-economic modelling of backhaul. This approach allows marketing teams to shout about high data rates while applying a pragmatic engineering solution to keep costs manageable.

3.6 Integrating 5G within a Multi-RAT Backhaul Network

Once the dimensioning model has been completed for all RATs, the required capacity will need to be mapped to an available access circuit capability. The 3GPP transport network layer for 5G is IP; typically this is still implemented as an IPv4 network, although in many networks the user data, sitting above the TNL, is increasingly IPv6, translated to IPv4 on the SGi (or N6) interface if required by the host server. The TNL of LTE is IP, while most networks have now migrated UMTS away from ATM to an IP TNL as defined in 3GPP Release 5 [9]. 2G Abis has never been specified as an IP interface within the standards, however all vendors now support a proprietary IP Abis interface, therefore all RATs can be supported by IP over Carrier Ethernet [1], often simply referred to as Ethernet (however it is not simple LAN Ethernet).

Figure 3.5 illustrates a multi-RAT network with 2G GSM, 3G UMTS, 4G LTE and 5G base stations co-deployed to a cell site connected to the aggregation network by optical fibre and a second cell site, subtended from the fibre hub, via a point-to-point millimetre-wave radio system, typically an E-band (71–76 GHz paired with 81–86 GHz) system. While the radio base stations are illustrated as individual entities, in reality many of these will be combined on a single-RAN/multi-RAT base station, depending on actual implementation. The main difference from the backhaul solution in Figure 3.1 for pre-5G deployment and this illustrated in Figure 3.5 for 5G in a multi-RAT environment is capacity scalability. This has occurred throughout the backhaul network, starting with the subtended site to the left, the microwave radio link has been replaced with an E-band millimetre-wave radio. E-band radio has a lower cost per bit than typical microwave radio systems and offers significant scalability, up to 20 Gbps depending on available spectrum and local regulations. There is, however, a capex cost to purchase the new radio and complete the planning and installation work; there may also be a cost for the spectrum, and this could be an annual licence fee payable to the local regulator. E-band spectrum regulations vary greatly throughout the world. The actual capacity delivered by the E-band radio will vary based on the selected radio channel bandwidth, the modulation scheme and associated RF power budget.

The fibre hub site has been upgraded to 10 Gbps of backhaul on an optical fibre link to the RAN aggregation site. Given this jump from what is typically a 1 Gbps circuit for an LTE site to 10 Gbps for 5G, the chances are that there will be lots of headroom above the calculated backhaul capacity requirements from Section 3.5. In this example though, this 10 Gbps circuit must support the local traffic and that from the subtended site. When calculating the backhaul capacity requirements of such a hub site, it is possible to add an overbooking factor to account for a level of statistical multiplexing gain. To support the uplift from 1 Gbps to 10 Gbps, the optical networking equipment (OLTE) has been swapped out at either end of the fibre; this will incur additional capex and likely drive higher opex too, depending on whether this is a leased circuit or self-owned/deployed. Even in the latter case there is a possibility of an opex increase as the cost of annual technical support is generally a percentage of the equipment cost.

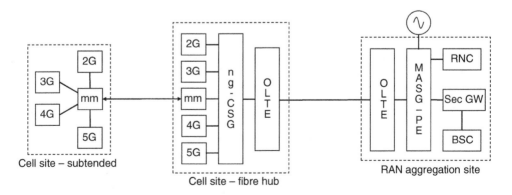

Figure 3.5   5G D-RAN mobile backhaul architecture.

The CSG in Figure 3.5 is labelled 'ng-CSG' as in next-generation, to differentiate it from the CSG in Figure 3.1, which would have been on site until the 5G upgrade is completed. The ng-CSG will have 10GE (Gigabit Ethernet) client ports and line ports along with support for 1GE as required. The ng-CSG will also support the necessary synchronization features to ensure frequency and phase (time of day) sync can be managed to meet the various radio interface requirements. For local EN-DC operation it is possible to establish the X2 interface between eNB and gNB via a local connection through the CSG, if co-located LTE and 5G are operating from different baseband units.

The network topology illustrated in Figures 3.1 and 3.5 is not uncommon, however in many cases it is likely that there are more intermediate hops between the cell site and IP Security Gateway, this probably involves the access circuit terminating at an IP/MPLS PE router with onwards transmission being provided via an IP VPN service, transiting core P routers as necessary. The use of IP VPN within the backhaul network enables many access circuits to be aggregated together, while IP/MPLS and VPNs in general offer opportunities to support a wide range of traffic types over a common IP network and underlying transmission, likely optical DWDM-based. This common IP core could support fixed and mobile traffic serving consumers and enterprise use cases. The flexibility of IP/MPLS enables traffic separation in different VPNs, resilient network routing through primary and secondary label switched paths, and massive scalability through high-speed access and network-facing interfaces. The underlying DWDM infrastructure is generally built as a series of point-to-point links with the network intelligence coming from the IP/MPLS layer.

Where such intermediate nodes are used it is essential to consider the economic implications of adding large numbers of 10GE access ports and the subsequent need for 100GE network-facing interfaces. In some cases, this may be a simple investment in more interface cards, not forgetting the upgrade to the core DWDM layer too, however this may drive the need for completely new routers along with the costs of introducing new platforms to the network. Even the relatively simple topology illustrated in Figure 3.5 has an MASG in the RAN aggregation site; in reality this site would likely contain a minimum of 2×MASG/PE to provide a level of resilience. The resilience could be 1+1 dual homing of all access circuits across both MASGs; this would offer high availability at a cost, alternatively adjacent sites could be mapped to alternative MASGs to minimize the impact of a failure of one of the two MASG platforms. The term 'MASG' originated with the development of pseudowire solutions, however it is used here to represent an aggregation platform which may be L2 and/or L3, including an IP router, depending on the specifics of a given network design. Similar resilience rules are common for the practical implementation of the IPsec GWs, given the size of geographical outage which would occur if an IPsec GW failed. The cost of doubling up on these platforms will be significant; an operator must consider the marketing implications of such a failure to build a robust financial case for investing in resilience, and how close to the cell site that resilience reaches. The cell site may be connected by a single fibre path or could be integrated on an optical fibre ring with primary and secondary (diverse) routes going different ways around the ring. Alternative approaches to backhaul resilience include the use of microwave or millimetre-wave radio to provide an alternative route to the primary path on fibre, or even the use of satellite communications, an option that's being used increasingly by mobile network operators to provide resilience for critical services. The wireless back-up solution, be this terrestrial or satellite, may not support as much capacity as the fibre connection, however it would keep some services up and running, with Quality of Service (QoS) mechanisms mapped to 3GPP QCIs (known as 5QI for 5G NR) being used to ensure availability of the resilient connection for certain traffic types in the event of congestion.

3.7 Use Case – BT/EE 5G Network in the UK

To conclude the specific focus on 5G D-RAN implementation, let's review a use case from BT in the UK. BT owns the UK mobile network operator EE, which launched a commercial 5G service on 30 May 2019. The EE 5G launch network was based on an EN-DC architecture and focused on addressing eMBB use cases in the first instance. The decision was taken to launch with a 5G D-RAN solution, however a level of flexibility was required to enable future RAN functional decomposition if required. This could include the implementation of an F1 interface for a centralized CU or the implementation of CPRI and/or eCPRI interfaces in support of a centralized DU and CU. The process described above was followed; a network capacity model was created, availability and performance calculations completed, and a techno-economic case prepared to justify the proposed mobile backhaul strategy and target architecture.

The EE mobile network benefits from a level of network sharing with another UK mobile network operator via a joint venture company known as MBNL (Mobile Broadband Network Limited). 3G UMTS is an active network share implemented as a MORAN solution; 2G GSM, 4G LTE and 5G are unilateral deployments, however in many cases the physical site and structure are shared along with power supply – in some cases passive antennas and backhaul are shared. The underlying driver is to find as many financial synergies as possible from sharing, while allowing both operators to implement their individual business and marketing strategies. These considerations must be factored into any techno-economic modelling.

EE decided to adopt a fibre-first strategy for 5G backhaul deployment, a change from what was previously a preference for microwave radio backhaul wherever possible. The move to an increasing volume of fibre backhaul started with the introduction of HSPA technologies and continued with the rollout of LTE. The primary driver is the increased capacity requirements of advanced multi-RAT cell sites, particularly when backhaul is shared between two network operators. This limits the practical implementation of certain microwave radio topologies and requires a greater fibre density to add distributed aggregation nodes for sites which are subtended via microwave radio links, similar to the topologies highlighted in Figures 3.1 and 3.5, where a single wireless backhauled site connects to a fibre-fed cell site. Other topologies include the addition of more links from the fibre hub in a star topology and/or chains of links extending from a hub site. The latter is increasingly challenging given the per site and link capacity demands.

One of the challenges of backhaul sharing is the ability of each operator to set their own end-to-end QoS models and change these as required to support new products and propositions. One way to ensure this flexibility is to share at the optical wavelength level rather than at an active Ethernet/IP layer. Access network-specific CWDM and DWDM products are available; the importance of access DWDM is that it can easily be amplified to extend the reach of the access circuit, it also offers significant scalability. EE selected an access DWDM solution as the primary fibre backhaul solution for 5G rollout to dense urban, urban and suburban areas. The solution is provided by Openreach, the UKs open whole-sale fixed access network provider and is known as OSA-FC; this refers to Optical Spectrum Access – Filter Connect [10]. The technical solution comprises two 1U boxes (1 rack unit = 44.45 mm in height), an active NTU and a passive filter module; in the case of EE a 16-channel filter module has been selected.

Figure 3.6 illustrates the BT/EE mobile backhaul network architecture. 2G GSM is supported on a multi-RAT/single-RAN base station, which also supports 4G LTE. The GSM/LTE base station typically connects to the CSG via a 1 Gbps Ethernet interface, however in the case of a high-capacity Gigabit Class LTE base station this connection is a 10 Gbps Ethernet interface. The Gigabit Class LTE service offers peak data rates of up to 1 Gbps and therefore requires greater than 1 Gbps of backhaul. The 10 Gbps interface is the next order of granularity, however the actual backhaul VLAN is shaped to 2 Gbps on the PE router. In co-transmission mode the multi-mode base station offers a single physical interface which supports Abis, S1 and X2 interfaces. The 3G UMTS service, as discussed above, is MORAN and therefore supports two network operators. Historically the UMTS MORAN supported significant capacity with up to 7×5 MHz FDD WCDMA carriers per cell sector; nowadays most of the carriers have been refarmed for use as LTE 2100. The Iub interface connection from the NodeB to the CSG is typically a 1 Gbps Ethernet interface, however nowadays a 100 Mbps connection would suffice. The 5G base station is added to the existing cell site and connected to the CSG via a 10 Gbps Ethernet interface to support the S1 (5G in EN-DC mode of operation and therefore connected to an EPC+ at this stage) and X2 interface. The local X2 interface between the co-located eNB and gNB is managed on site as a direct connection through the CSG; IPsec is implemented on this interface to provide authentication of connected network elements.

An alternative to the multiple base-station model presented in this case study would be a swap-out for a new multi-RAT solution which could support all radio access technologies, including 5G. This would be expensive and is likely to be uneconomical unless other factors drive a swap-out programme; these other factors could include network integration

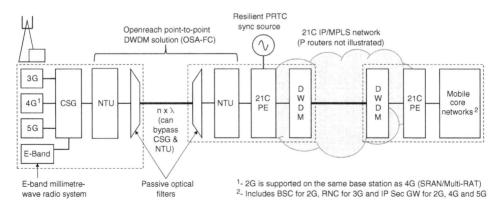

E-band millimetre-wave radio system   Passive optical filters   [1]- 2G is supported on the same base station as 4G (SRAN/Multi-RAT)
[2]- Includes BSC for 2G, RNC for 3G and IP Sec GW for 2G, 4G and 5G

Figure 3.6 BT/EE multi-RAT high-level mobile network architecture.

after the merger of two operators or an optimization of site design if physical accommodation issues were preventing further network growth. In reality, many network operators will add 5G as a new base station and interconnect this with the existing LTE base station by an X2 interface.

Figure 3.6 includes an E-band radio system; this is shown for completeness, however it wouldn't be deployed on every site. The vast majority of sites within the initial 5G rollout plan are urban and therefore the probability of connecting the sites on fibre is high. E-band millimetre-wave radio will be used to extend connectivity to any sites which can't be connected by fibre. The E-band radio connects to the CSG by a 10 Gbps Ethernet interface. The actual date rate of the E-band radio system depends on the RF configuration of channel bandwidth and modulation scheme.

The CSG connects to an active optical network termination unit; this provides a standard 'grey' optical interface to the CSG and a coloured optical interface (ITU-T DWDM grid [11]) towards the optical filter. The use of an active NTE here is primarily driven by the regulatory environment in the UK; such regulatory considerations will be an important part of the overall techno-economic analysis. The coloured 10 Gbps connection to channel 1 of the 16-channel filter module leaves another 15 channels available for future capacity upgrades, transmission sharing and/or RAN evolution. This last point will be explored further in the sections on C-RAN. From a capacity perspective, if the E-band radio traffic loading on a given fibre-connected site became greater than that which could be accommodated within the single 10 Gbps circuit, the E-band radio could connect directly to channel 2 of the optical filter by inserting a coloured optical transceiver on the traffic data port of the radio unit, or using another 'grey' port/module on the active NTE – the relative costs would determine which approach was optimal.

If dark fibre is available then alternative backhaul options are possible; the CSG could light the dark fibre directly from a pluggable optical transceiver, removing the need for the active NTE. The DWDM filter may or may not be required with dark fibre; the relative cost of multiple fibre cables should be compared with the cost of the DWDM filters and coloured optical transceivers. The relative economics of these different approaches will vary significantly between different markets and regulatory regimes.

The backhaul transits the access DWDM product and terminates within a RAN aggregation site on a PE router (MASG role). This PE router within the BT network serves as a multi-service edge router as, along with RAN backhaul aggregation, it aggregates fixed business and broadband traffic, acting as a broadband network gateway for the latter. Using high-end PE routers in this multi-service mode will reduce the overall total cost of ownership by maximizing the use of available network assets. Due to the likely high volume of 10GE access interfaces/ports, the network-facing interfaces of the PE router will need to be operating at 100 Gbps or higher; such high-speed optics are expensive and therefore careful consideration should be given to network traffic engineering, backed by an understanding of achievable statistical multiplexing gains.

The PE router also acts as the ingress points for network timing in the form of frequency and phase synchronization signals. The synchronization inputs come from a Primary Reference Timing Clock (PRTC), typically a highly accurate enhanced PRTC (ePRTC) platform. This platform provides frequency synchronization references which can be distributed directly over SyncE [2, 3, 12] or via Precision Timing Protocol (PTP) [13] packets.

Frequency synchronization is a common requirement for radio systems and is used to provide a stable frequency reference to discipline the base station oscillator for FDD and TDD operation. Phase (or time of day) synchronization is distributed via PTP packets and is used in some advanced LTE radio schemes such as enhanced Inter-Cell Interference Coordination (eICIC) and Coordinated Multipoint Transmission and Receptions (CoMP), however it is the use of TDD in 5G which is really driving the significant uptake in phase synchronization. TDD operation requires very accurate timing alignment between adjacent base stations to ensure switching between downlink and uplink transmission happens at the same instance, to avoid radio interference. The most common source of time synchronization is from Global Navigation Satellite Systems (GNSS). A local cell-site GNSS solution would work, however this would be susceptible to local jamming and spoofing. A network-based synchronization signal, based on distributed GNSS sources, offers increased network availability. These timing signals (frequency and phase/time of day sync) will be delivered to the base station by SyncE (frequency sync only) and/or IEEE 1588-2008 (frequency and phase sync). The most robust delivery mechanism is to utilize synchronous Ethernet and IEEE 1588-2008 on the access circuit. The benefit this brings is the ability for SyncE to provide an aid to local hold-over if the phase sync signal is lost, therefore enhancing the duration of hold-over and maintaining full operational status.

The 100GE network-facing interfaces connect to transponders on the active core DWDM system, which provides high-speed and high-capacity connectivity with the core P routers, these high-end label-switching routers will require significant throughput. The mobile backhaul IP VPNs transit this high-speed IP/MPLS core transport network to the centralized mobile core network. The term 'centralized' is of course relative; this will differ from country to country based on the geographical footprint of a given network operator. In the case of BT's mobile network, the core network resides at 12 geographical distributed nodes, so while referred to as centralized, it is in fact distributed to geographical regions to optimize network costs while enhancing user experience. These mobile core network hubs accommodate the IP Security Gateways along with BSC and RNC elements (accommodating some of the network elements which previously resided on RAN aggregation sites). Additionally, these sites accommodate the traditional mobile network core nodes in support of circuit- and packet-switching functions along with associated databases, policy engines, security and optimization capabilities.

The mobile core network hubs also provide access to Content Distribution Networks (CDNs), which provide direct local access to a range of popular video content, therefore saving on transmission costs to Internet peering/transit locations and associated internetwork charges. The business case for CDNs isn't straightforward though. Firstly an operator needs to decide whether a third party will provide the physical hardware or whether this will be provided by the operator. Quite often the CDN solution is installed by the third-party content provider, however the operator will be liable for power consumption and managing any heating, ventilation and air-conditioning requirements to provide the correct environmental conditions for the CDN servers. A techno-economic analysis is required to understand the value of deploying CDNs and the number of locations in which CDNs make commercial sense. Recent developments in online streaming gaming services means that such CDN locations will become broader service platforms, which will address a range of latency-sensitive services. An operator must develop a clear strategy on how they will interact with content providers in this space. An example of this is the partnership between BT and Google, in which Google's Stadia online gaming service is hosted within the BT core hub sites [14].

3.8 5G C-RAN – F1 Interface and Midhaul

C-RAN means different things to different people, from Cloud RAN to Coordinated RAN to Cooperative RAN – all are effectively the same thing, however the level of coordination or cooperation depends on where the functional split takes place. Figure 3.3 illustrates the functional decomposition of the 5G gNB into RU, DU and CU components. This section reviews the 3GPP-defined option 2 split, which is supported via the standardized F1 interface [8, 15–19]. The option 2 split is a higher-layer split within the gNB protocol architecture and sits between the PDCP layer and high RLC. Figure 3.7 illustrates the position of the option 2 split and the F1 interface; this diagram is based on 3GPP TR 38.801 [20]. The diagram is applicable to a 5G EN-DC network, however the only thing which changes once an operator migrates to a 5G core network is the inclusion of a new protocol between the user data (data box to left of diagram) and PDCP. This new function is the Service Data Adaptation Protocol (SDAP), which has been introduced in the NR user plane to handle the flow-based QoS framework in RAN, such as mapping between QoS flow and a data radio bearer, and QoS flow ID marking.

Given that the F1 interface enables a higher-layer split, this limits the amount of coordination benefit which can be achieved as the higher-layer split occurs within the non-real-time portion of the protocol stack. That said, there are other benefits to be had from a DU–CU split; the CU contain the PDCP layer which, amongst other things, supports encryption and decryption of the control and user plane traffic on the radio interface. Moving this function to a trusted domain within the operator's network offers an option to remove IPsec ESP (Encrypted Security Payload) from the user plane traffic over the F1

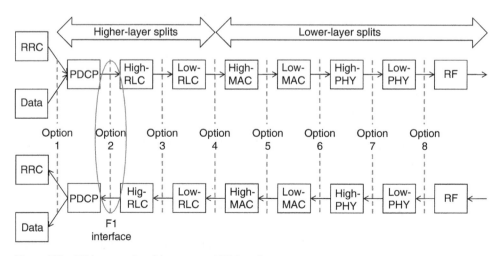

Figure 3.7 RAN protocol architecture and F1 interface.

interface. This not only reduces the transport network overheads, but also provides a finer granularity of hashing (load-balancing) which enables a more optimal utilization of back-haul links, particularly if aggregation links deeper in the network are operating at $n \times 10$ Gbps. This could have a significant impact on the transport network costs by delaying or removing the need to upgrade $n \times 10$ Gbps inter-router links to 100 Gbps links. An additional benefit of moving the PDCP layer to a trusted site is security; the control and user plane data is now fully encrypted from the UE to a trusted node (CU) within the mobile network operator's infrastructure, rather than requiring the radio traffic to be decrypted at the cell site and then re-encrypted by IPsec for onwards transmission towards the core.

Industry initiatives such as O-RAN [21] propose the addition of a new node to the 5G network architecture; this new node is known as a RAN Intelligent Controller (RIC) and operates on the management plane. The RIC could utilize machine learning to provide radio network optimization across a number of DU/RU combinations which are connected to a common CU. It should be noted that the RIC can also support radio optimization, even with the distributed CU function as found in a D-RAN – albeit this will probably be less efficient. O-RAN specifies two types of RIC, a near real-time RIC which will run xApps and a non-real-time RIC which will run rApps. The use of software applications enables ongoing network and service optimization.

The transport network connectivity requirements of an F1 interface are sometimes referred to as 'midhaul', however it is not uncommon to hear this interface referred to as 'backhaul', so it's always best to clarify this point when discussing RAN functional splits and architecture decomposition. The dimensioning of the F1 midhaul interface is not dissimilar to that of the S1 backhaul network of an EN-DC architecture; this itself is very similar to the dimensioning of backhaul to a next-generation core network over the N2/N3 interfaces. An analysis of the F1 interface dimensioning shows a typical uplift from S1 backhaul of approximately 5%; this is likely to be within the error margin of the S1 interface dimensioning model. The functions of the CU can easily be accommodated within a general computing platform with no need for any specialized hardware acceleration; as such, the CU can be implemented as a software function on a network cloud instance, as a Virtual Network Function (VNF).

Given the similarities in data rates between backhaul and F1-based midhaul transmission, the same set of technical solutions are valid. An option 2 split introduces no RAN-specific latency objectives and therefore, as with backhaul, the allowable latency budget is based on the service offerings the operator wishes to productize and market. Midhaul can therefore make use of any optical or wireless transport products – such as point-to-point fibre solutions – including access DWDM as discussed above, or Carrier Ethernet services. Likewise, any wireless solution which is suitable for backhaul may also be used for mid-haul; these include traditional microwave radio links and millimetre-wave radio systems. Both can be implemented as point-to-point links along with point-to-multipoint and mul-tipoint-to-multipoint (mesh) systems.

3.9   5G C-RAN – CPRI, eCPRI and Fronthaul

A truly coordinated RAN is realized when a lower-layer split is implemented. This enables real-time coordination between distributed RUs connected to a common DU function. In this case the CU may be collocated with the DU or located at an even more centralized location; this is of no relevance to the fronthaul discussion. The reference to typical CPRI data rates of 4G LTE C-RAN in Section 3.3 illustrates the significant transmission capacity required for an option 8 split, while Section 3.4 introduced the concept of eCPRI, an option 7 split. It is this eCPRI split which will support 5G massive MIMO implementations, although, as discussed, there will be many lower-order MIMO systems which are implemented using CPRI with proprietary compression algorithms. Massive MIMO systems of 16T16R and greater are already implemented with an option 7 eCPRI split, and this is the direction of development for many future 5G systems. 3GPP specifies three variants of the option 7 split, these are known as 7-1, 7-2 and 7-3 (note that 7-3 is downlink only). In addition to the 3GPP splits, O-RAN Alliance has defined an eCPRI-like split known as 7-2x (Open Fronthaul). All these option 7 variants can be carried over an Ethernet-based fronthaul.

While an option 7 split reduces the data rate required on the DU–RU interface, it doesn't change the tight latency requirements of 250 μs; this is based on the HARQ process, which sits in the lower MAC layer of the 5G NR protocol stack. The 250 μs time refers to the allowable round-trip time (there and back plus processing time) and as such limits the typical one-way fibre distance of ~20 km. Different vendors provide different optimizations on this interface, so the exact maximum distance may vary slightly between vendors – however most of the delay budget is used for physical transmission over the optical fibre cable. Lightwave transmission travels at approximately two-thirds of the speed of light when contained within a single-mode optical fibre strand. The location of the option 7 eCPRI split is highlighted in Figure 3.8. It should be noted that O-RAN [21] can support non-ideal fronthaul, which refers to fronthaul which doesn't meet the tight 250 μs RTT. If the latency exceeds the 250 μs RTT it is reasonable to expect that some of the benefits of a lower-layer split might not be realized in the presence of non-ideal transport. This will particularly impact the real-time coordination between radio units.

An option 7 split results in the implementation of the lower physical layer in the RU, be this a standalone radio unit or an integrated function within an active/massive MIMO antenna. This results in a more complex technical solution as effectively the real-time components of the baseband unit are being split across two network elements, the RU and DU.

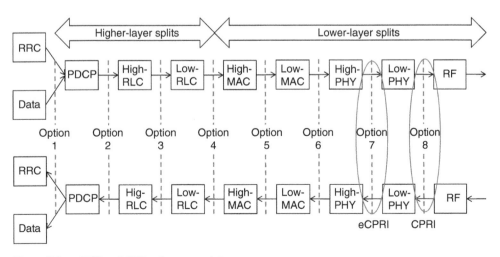

Figure 3.8 eCPRI and CPRI reference points.

While the functional split has moved from option 8 to option 7, the DU still contains real-time functionality and as such can't easily be virtualized on common compute platforms in the way the non-real-time CU can. This results in the need for specialist hardware to support the high physical layer functionality. Today this is realized by dedicated hardware solutions from the major RAN vendors, however there is a move towards a combined common compute platform with a suitable accelerator module which simply plugs in to the compute chassis. This may ultimately change the economics of building macro-cell radio networks. Longer term we will likely witness a significant evolution in computing power such that dedicated hardware isn't required. For certain small-cell applications there are already vendors who can demonstrate a DU with an option 7 split running on commercial compute platforms, however these solutions are very different from the high-end high-capacity solutions required for wide-area macro-cellular coverage.

The implementation of a functional split within the RAN may vary between RATs and even frequency bands within a RAT. If different splits are implemented for different bands within a RAT, it is vital to consider the implications this will have for the use of Carrier Aggregation (CA). Figure 3.9 illustrates the range of gNB architectures which can be implemented due to the flexibility introduced through the functional decomposition of the RAN. For the purpose of reviewing the options, they have been labelled 1, 2, 3 and 4 in Figure 3.9. Implementation number 1 is a fully distributed (also referred to as fully aggregated) gNB; this is as described in the use case in Section 3.7. Implementation number 2 illustrates an option 2 split with an F1 interface in which the DU and CU are physically separated, however the RU(s) are collocated with the DU; this solution is reviewed in Section 3.8. Implementation number 3 is the first of two solutions for a real-time coordinated RAN deployment in which the RU(s) and DU are installed on physically separate locations (sites). The diagram lists CPRI or eCPRI as the fronthaul interface, however in the case of

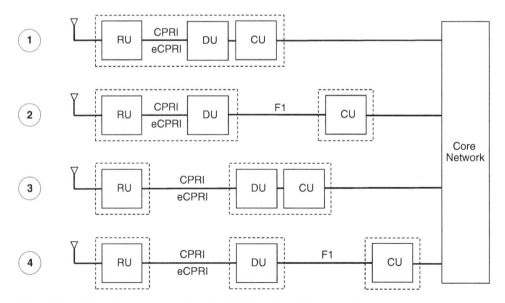

Figure 3.9 gNB architectures – options for RAN functional decomposition.

an O-RAN implementation this would be an Open Fronthaul interface. In this solution the DU and CU are collocated. Implementation number 4 builds on number 3 by separating the DU and CU; this would enable the CU to provide a common point of control and optimization for multiple DUs. This optimization could take the form of a co-located RAN Intelligent Controller as defined within O-RAN specifications.

Note that Figure 3.9 assumes no separation of CU control and user plane as illustrated in Figure 3.4. If the E1 interface is implemented, then additional architectures would be possible with the CU-c located in a different location to the CU-u. A single scaled CU-c instance could support multiple subtended CU-u network elements.

3.10 Connectivity Solutions for Fronthaul

A cell site deployed in a C-RAN configuration with fronthaul connectivity between local RUs and a centralized DU requires a transmission/transport network solution which can support the high data rate and low latency requirements of CPRI and/or eCPRI. A techno-economic analysis is required in the first instance to justify the implementation of a fronthaul network; therefore, the benefits of a C-RAN will need to be proven for a given operator. The techno-economic analysis will use the existing D-RAN deployment costs as the baseline. Advantages of a C-RAN will vary between different markets, however factors to consider include reduction of equipment at cell site (possibly leading to lower site rental costs), reduced opex from having all baseband in a subset of locations, lower electrical power consumption at the cell site, statistical multiplexing gains from pooling of baseband digital signal processing resources (which reduces the overall baseband requirements) and improved radio network performance, including gains in spectral efficiency. In addition to the financial benefits of a C-RAN, an operator must consider the costs associated with building and operating the new RAN architecture. Such considerations include provision of suitable transmission systems for fronthaul, accommodation and power for baseband units, any generic compute requirements (particularly if the CU is to be accommodated deeper in the network than the DU) and suitable frequency and phase/time of day synchronization to support the RAN requirements.

Once the strategic decision to implement a C-RAN has been taken, the operator will need to develop the transport network strategy and target architecture for the fronthaul network. Once this is in place, high-level and low-level designs can be produced and network planning can commence. A radio interface capacity model will inform the development of the technical fronthaul solution and scalability requirements of the baseband DU and CU functions. The latency constraints of the HARQ process (a function of the MAC layer) will be a key input to the design of the fronthaul network. An eCPRI interface circuit can be realized over an optical fibre or wireless connection, or even a hybrid of the two. In addition to these two established methods of connectivity, a TIP (Telecom Infra Project) on vRAN Fronthaul is currently ongoing to establish mechanisms for transmitting fronthaul over non-ideal connectivity; this refers to systems which don't have the required throughput or exceed the latency of a conventional fronthaul solution [22]. Due to the significantly higher capacity demand of a CPRI interface, compared with eCPRI, it is more challenging to realize CPRI solutions on wireless fronthaul, albeit this can be achieved for a low carrier and antenna count.

Over the remainder of this section, we'll explore the technical and economic consider-ations for the choice of optical and/or wireless fronthaul solutions. Fibre-based optical net-works involve a significant upfront cost for digging trenches, installing cable ducts, pulling fibre-optic cable through the ducts and splicing. A completely new build is unlikely to cost in for C-RAN fronthaul only. However, it may do so if several use cases are combined. For example, mobile backhaul, midhaul and fronthaul along fixed broadband fibre to the cabinet/premise plus business Ethernet and IP VPN services. An alternative to fibre self-build is to lease dark fibre from the incumbent operators or established competitive fibre provider. In some countries dark fibre is available at low cost; this is particularly common in Asia and has resulted in widespread adoption of C-RAN in urban and suburban areas. In other markets dark fibre is either expensive or not available at all. An alternative to dark fibre is passive infrastructure access; in this case the regulator may mandate that the incum-bent operator allows access to its fibre ducts and, where applicable, telephone poles too – those typically being used for local drop wires. In the case of such passive infrastructure access, a mobile operator would install their own fibre-optic cables in existing ducts and pay an agreed annual rental fee (an opex) for the continued use of the ducts. The fibre options we've reviewed above are all passive solutions; a fibre is provided which will require the addition of optical networking equipment to light the service. An alternative to a passive fibre service is an active service – the delivery of an optical networking solution such as the Openreach OSA-FC solution described in Section 3.7. Other lit services may provide high-speed Carrier Ethernet services which can support the direct connectivity of eCPRI while also supporting any other midhaul or backhaul requirements from the cell site. Recent developments in Radio over Ethernet technologies based on IEEE 1914.3 offer an alternative approach to DWDM for CPRI fronthaul; this also facilitates the aggregation of CPRI, eCPRI/Open Fronthaul, midhaul and backhaul on a common Carrier Ethernet bearer.

Irrespective of whether the access connectivity for fronthaul is a lit or dark service (active or passive), someone has to provide the optical networking equipment to enable the circuit to operate. The mobile operator needs to understand the range of options available as these will inform the overall techno-economic analysis, a fundamental part of which is whether to self-build – via one of the methods discussed – or lease a connectivity service. The sim-plest technical solution is to deploy point-to-point fibre cables between the cell site and the DU baseband location. Most macro-cell sites consist of three cell sectors, albeit sites with more or less cell sectors are in use too; each cell sector will be provisioned with one or more radio units and/or active antennas. Each radio unit will require a physical fronthaul con-nection to support the CPRI or eCPRI interface; there are several options available for the network connectivity of this interface between the cell site and centralized DU. The local fibre to the tower connectivity within the cell site will be either a multi-mode or single-mode fibre pair, which connects to the remote DU.

The network connectivity could consist of a pair of fibre-optic cables for each radio module on each cell sector; this would result in the need for a considerable amount of fibre, though the cost of this may be prohibitive while the cable management at the centralized DU loca-tion will be complex. Alternatively, there are active and passive solutions which could reduce the quantity of fibre cables required for the fronthaul network. Active solutions are based on CWDM and DWDM technologies; both enable the conversion of standard 'grey' optics from the radio modules to 'coloured' optics which align with a coarse or dense wavelength grid. Additionally, the active equipment will provide power amplification to drive the fronthaul fibre transmission. xWDM fibre fronthaul systems were typically based on dual-fibre working, however an increasing range of single-fibre working systems are becoming available for access network applications. A single-fibre working solution is constrained on the number of channels supported, as a bi-directional channel requires two wavelengths whereas only one wavelength is required on each of the two fibres in a dual-fibre solution. A passive optical filter could be used as an alternative to an active xWDM system. In this scenario, the pluggable optical transceivers in the radio module would have to be channel specific. Such coloured optics are more expensive than standard grey optics. Another challenge with the use of coloured optics is the quantity of different transceivers that an operator needs to purchase for rollout and spares. Programmable coloured optics offer a solution to this challenge, however such modules tend not to be temperature hardened and/or of a sufficiently high output power to drive the fibre distance to the centralized DU. Optical transceiver manufacturers are working to address both of these issues, however once they do, the temperature-hardened transceivers are likely to be expensive – at least until significant volumes start to ship. In the future, this will certainly be a viable solution which will simplify installation and minimize the costs associated with spares holding. Recent trends in access network connectivity, including mobile backhaul, have driven significant adoption of 10 Gbps Ethernet services. As this trend continues and 100 Gbps Ethernet access becomes mainstream, it will enable Radio over Ethernet implementation and position this technology as a flexible and potentially lower-cost alternative to xWDM fronthaul. Mobile network operators should conduct a techno-economic analysis to inform their decision-making, taking into account the interface type used to enable the fronthaul connections to the DU. This may evolve from $n \times 10$ Gbps and 25 Gbps to $n \times 100$ Gbps as virtualized DU is deployed.

Fronthaul over wireless was constrained but not impossible with an option 8 CPRI interface, however an option 7 eCPRI interface – with the reduced capacity requirement – makes wireless solutions much more viable. The exacting latency requirement from the HARQ process is still a consideration, however a signal travels faster over the air than it does through a fibre-optic cable, therefore wireless has a role to play in fronthaul connectivity. The recent developments in millimetre-wave radio systems, driven by ETSI ISG mWT [23], have established a competitive range of suppliers who offer high-capacity E-band radio solutions which can support multiple fronthaul circuits for a site with $n \times$ carriers per cell sector. Further innovations in W-band and D-band radio systems will ensure that enough radio carriers of the required RF channel bandwidths can be supported in an urban environment. E-band radios operate in the frequency bands of 71–76 GHz and 81–86 GHz; W-band is 92–114.5 GHz while D-band is 130–174.5 GHz.

For geographical regions which require greater physical reach than the link lengths achievable with millimetre-wave radio systems, advances in the traditional microwave radio bands of 6–42 GHz will help. Wider RF channels, higher-order modulation schemes and higher transmit output power will all increase system throughput, however the first two enhancements will impact system gain and therefore shorten the link length for a given configuration. Further wireless considerations should include the recent advances in dual-band operation in which a combination of microwave plus millimetre-wave radio channels is aggregated or the development of carrier aggregation capabilities within the traditional microwave radio bands, either intra- or interband.

Fronthaul over non-ideal backhaul [22] may also have a role to play; research is ongoing to understand the play-offs between ideal fronthaul and the performance of the radio interface and how this changes as factors such as throughput, latency and packet error loss rate vary. The research is exploring a range of physical transmission systems such as DOCSIS, G.Fast, Structured Ethernet cabling for LAN and wireless solutions.

3.11 Small Cells in FR1 and FR2

3GPP has defined two frequency bands for 5G NR – most, although not necessarily all, macro cells will operate in FR1; this refers to the spectrum between 410 MHz and 7.125 GHz. Small cells will also operate in FR1, however the higher FR2 will also be deployed in the small cell layer. FR2 refers to the spectrum between 24.250 GHz and 52.600 GHz (WRC19 has specified bands above the current upper limit). Early indications from vendor roadmaps suggest that in many cases, the FR1 5G small cell will be a radio unit with a CPRI interface to connect to a remote DU, to enable coordination and optimization of the FR1 transmissions within the dense small-cell layer. FR2 product roadmaps suggest a combined RU and DU module with an F1 interface which will connect to a remote CU. These different approaches will influence the choice of connectivity solutions which can support the two different frequency ranges. BT has recently deployed a 5G small-cell trial with a C-RAN architecture; this is operating in EN-DC mode with all radio units, LTE and NR, being connected with CPRI interfaces and configured for 4T4R mode of operation. eCPRI for such radio units is starting to appear on roadmaps and will start to ship in volume as O-RAN-based small cells are deployed with eCPRI/Open Fronthaul as standard.

Small-cell topology is a key consideration when selecting the transmission solutions for small-cell deployments. Options include subtending the small cells from an existing macro cell, connecting directly to a fibre network along with hybrid fibre and wireless topologies in which small cells are subtended from other small-cell sites, not dissimilar to macro-cell backhaul topologies which were reviewed earlier.

The backhaul, midhaul and fronthaul transmission solutions which have been reviewed within this chapter in the context of macro-cell connectivity are all applicable to small cells, albeit different form factors may be required. The target price point of a small-cell solution is far more challenging given the limited coverage footprint; it is essential to minimize costs while maximizing spectral efficiency and therefore capacity, often the main driver for deploying a small cell. Radio systems operating in D-band are an exciting innovation, which will help manage the cost of small-cell deployments, likewise developments by 3GPP on Integrated Access and Backhaul (IAB) may offer opportunities for lower-cost network densification, particularly in FR2.

3.12 Summary

5G mobile backhaul, midhaul and fronthaul solutions contribute significantly to the capex and opex associated with rolling out the next generation radio access network. The choice of RAN architecture, and in particular the functional decomposition of the RAN, will influence the transport network design options available to an operator. The majority of early 5G deployments are based on the NSA EN-DC architecture and as such tend to mirror the 4G LTE RAN. In many markets the D-RAN architecture dominates, while in certain markets the commercial and regulatory environments relating to the availability of dark fibre have promoted the deployment of C-RAN. While fibre is the connectivity solution of choice where it is available and commercially viable, wireless connectivity solutions in the microwave and millimetre-wave bands offer a viable alternative. Development in 3GPP IAB should be followed closely; the potential implementation of the connectivity requirements for massive network densification is significant when used in the optimal architecture with fibre and fixed wireless links. The need to deliver frequency and time of day (phase) synchronization remains an essential function of the transport network. Key performance indicators such as latency, packet delay variation, packet error loss rate and availability are increasingly visible as society becomes ever more reliant on mobile communications. The continued use of CPRI for many 5G radio units, along with the adoption of dynamic spectrum sharing as a means of migrating capacity from LTE to NR, results in a mixed CPRI/eCPRI requirement for many networks for the foreseeable future. This, along with O-RAN Open Fronthaul, will need to coexist within fronthaul transport networks. The difference between the optimal techno-economic connectivity solution and the wrong one will have a significant impact on the performance and therefore profitability of a mobile network operator.

Acknowledgement

The author would like to thank Dr Richard MacKenzie of BT Applied Research for his review of the draft manuscript and constructive feedback which has improved the quality of this chapter.

References

1 Metro Ethernet Forum. (29 March 2020). MEF home page. Available at http://www.mef. net (accessed 29 March 2020).

2 International Telecommunications Union. (29 August 2019). G.8261: Timing and synchronization aspects in packet networks. Available at https://www.itu.int/rec/T-REC-G.8261-201908-I/en (accessed 29 March 2020).

3 International Telecommunications Union. (29 November 2018). G.8262: Timing characteristics of a synchronous equipment slave clock. Available at https://www.itu.int/rec/T-REC-G.8262-201811-I/en (accessed 29 March 2020).

4 CPRI Specifications Group. (9 October 2015). CPRI v7 common public radio interface (CPRI); Interface specification. CPRI Specifications Group. Available at http://www.cpri. info/downloads/CPRI_v_7_0_2015-10-09.pdf (accessed 29 March 2020).

5 NGMN Ltd. (31 March 2015). NGMN RAN evolution project – backhaul and fronthaul evolution. Available at https://www.ngmn.org/wp-content/uploads/NGMN_RANEV_D4_BH_FH_Evolution_V1.01.pdf (accessed 29 March 2020).

6 CPRI Specification Group. (22 August 2017). eCPRI v1 common public radio interface: ECPRI interface specification. Available at http://www.cpri.info/downloads/eCPRI_v_1_0_2017_08_22.pdf (accessed 29 March 2020).

7 CPRI Specification Group. (10 May 2019). eCPRI v2 common public radio interface: ECPRI interface specification. Available at http://www.cpri.info/downloads/eCPRI_v_2.0_2019_05_10c.pdf (accessed 29 March 2020).

8 3GPP. (11 December 2017). 3GPP NG-RAN; F1 general aspects and principles. Available at https://portal.3gpp.org/desktopmodules/Specifications/SpecificationDetails. aspx?specificationId=3257 (accessed 27 March 2020).

9 3GPP. (6 April 2001). 3GPP portal. Available at https://portal.3gpp.org/desktopmodules/ Specifications/SpecificationDetails.aspx?specificationId=1355 (accessed 29 March 2020).

10 British Telecommunications plc 2020. (3 April 2018). Openreach OSA-FC. Available at https://www.openreach.co.uk/orpg/home/updates/briefings/ethernetservicesbriefings/ ethernetservicesbriefingsarticles/eth01218.do (accessed 29 March 2020).

11 International Telecommunications Union. (13 February 2011). ITU-T recommendation G.694.1. Available at https://www.itu.int/rec/T-REC-G.694.1-201202-I/en (accessed 29 March 2020).

12 International Telecommunications Union. (29 August 2017). G.8264: Distribution of timing information through packet networks. Available at https://www.itu.int/rec/T-REC-G.8264-201708-I/en (accessed 29 March 2020).

13 IEEE SA. (27 March 2008). 1588-2008: IEEE standard for a precision clock synchronization protocol for networked measurement and control systems. Available at https://standards. ieee.org/standard/1588-2008.html (accessed 29 March 2020).

14 British Telecommunications plc. (17 January 2020). BT becomes first European network to partner with Google on new Stadia home broadband offers. Available at https://newsroom. bt.com/bt-becomes-first-european-network-to-partner-with-google-on-new-stadia-home-broadband-offers (accessed 29 March 2020).

15 3GPP. (11 December 2017). NG-RAN; F1 layer 1. Available at https://portal.3gpp.org/ desktopmodules/Specifications/SpecificationDetails.aspx?specificationId=3258 (accessed 29 March 2020).

16 3GPP. (11 December 2017). 3GPP NG-RAN; F1 signalling transport. Available at https:// portal.3gpp.org/desktopmodules/Specifications/SpecificationDetails.aspx?specificationId=3259 (accessed 29 March 2020).

17 3GPP. (11 December 2017). 3GPP NG-RAN; F1 application protocol (F1AP). Available at https:// portal.3gpp.org/desktopmodules/Specifications/SpecificationDetails.aspx?specificationId=3260 (accessed 29 March 2020).

18 3GPP. (11 December 2017). 3GPP NG-RAN; F1 data transport. Available at https:// portal.3gpp.org/desktopmodules/Specifications/SpecificationDetails. aspx?specificationId=3261 (accessed 29 March 2020).

19 3GPP. (11 December 2017). 3GPP NG-RAN; F1 interface user plane protocol. Available at https://portal.3gpp.org/desktopmodules/Specifications/SpecificationDetails. aspx?specificationId=3262 (accessed 29 March 2020).

20 3GPP. (6 March 2017). 3GPP study on new radio access technology: Radio access architecture and interfaces. Available at https://portal.3gpp.org/desktopmodules/ Specifications/SpecificationDetails.aspx?specificationId=3056 (accessed 29 March 2020).

21 O-RAN Alliance. (29 March 2020). O-RAN home page. Available at https://www.o-ran.org (accessed 29 March 2020).

22 Telecom Infra Project. (29 March 2020). TIP vRAN Fronthaul. Available at https:// telecominfraproject.com/vran (accessed 29 March 2020).

23 ETSI. (29 March 2020). ETSI industry specification group (ISG) on millimetre wave transmission (mWT). Available at https://www.etsi.org/committee/mwt (accessed 29 March 2020).

4

Key 5G Transport Requirements

Kenneth Y. Ho and Esa Metsälä

4.1 Transport Capacity

4.1.1 5G Radio Impacts to Transport

The new 5G spectrum at mid and high bands and the use of massive MIMO (Multiple Input Multiple Output) antennas increase the traffic-handling capacity of the network and mandate an upgrade to the network capacity for transport links and nodes also. Beamforming and massive MIMO extend particularly capacity peaks and make the difference between peak and average capacity for transport bigger in 5G than in LTE.

In FR1, the main objective with MIMO is to increase spectral efficiency as the bandwidth is limited. In FR2, the main objective with MIMO is to increase coverage as the attenuation is higher. The highest capacities for transport also come from higher-frequency bands like FR2, where the air interface bandwidth is the largest.

The notation $n \times m$ MIMO (e.g. 2×2 MIMO) refers to n transmit antennas and m receive antennas. With multiple transmit and receive antennas, the signal may use multiple, separate paths in the air interface. These air interface paths provide added capacity when different signals are sent over the different paths, and path diversity when the same signal is sent over multiple paths.

Directing the beam is achieved by using multiple antenna elements and shifting the signal phase for these elements, with the objective of constructive reception in the targeted receiver. An example antenna construction was shown in Figure 2.17 of Chapter 2, with 192 antenna elements, 64 TRXs and support for 16 streams.

With 5G multi-user MIMO, a sector supports multiple users each using multiple streams, maximizing capacity. The peak capacity of the sector depends on the amount of streams per user and the amount of users.

The maximum theoretical peak capacity is seldom achieved in practical networks as it would require ideal radio conditions for several users so that each user gets maximum bit rates simultaneously and independently from the other users. More realistically, there can still be sectors where there are few users statically located (e.g. not moving) close enough to the cell centre to get a maximum bit rate on multiple streams yet far enough from each other so as not to cause high interference. This is the case on the left-hand side of Figure 4.1,

5G Backhaul and Fronthaul, First Edition. Edited by Esa Markus Metsälä and Juha T. T. Salmelin.
© 2023 John Wiley & Sons Ltd. Published 2023 by John Wiley & Sons Ltd.

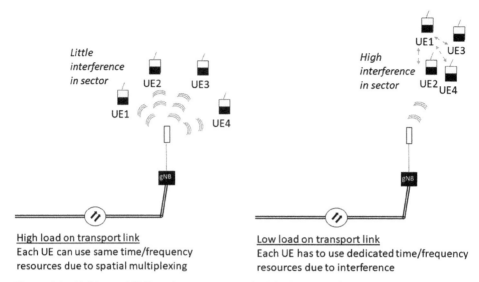

Figure 4.1 Multi-user MIMO and transport capacity (single sector shown).

where there are four UEs all receiving multiple streams in a condition of little interference from the other UEs, as these are spatially separated in such a way that there is little interference and all UEs are effectively isolated from each other. This situation could exist if, for example, buildings block the signals of the other UEs.

When each UE can reuse the same time and frequency resources, the throughput is much higher for the backhaul link.

Dimensioning the backhaul link from gNB is up to the dimensioning strategy to select a suitable transport line rate between the (theoretical) peak, taking into account massive MIMO and spatial multiplexing, and average throughput. Favouring the average value creates a higher probability of occasional packet losses on the transport links, while favouring the peak value leads to higher transport bandwidth needs and higher transport opex. On the backhaul, packet loss for mobile broadband-type background traffic causes degradation of user throughput and user experience.

Once the target value is found, the line rate is selected to be some of the common Ethernet line rates, like 10G, 25G, 50G or 100G. The interface rate may actually be higher than the true service rate, if the Ethernet service is provisioned for a rate less than the line rate. Shaping is required at the egress of the source node (gNB) to avoid service provider ingress node policer dropping packets.

For multiple radio access technology sites, like 5G with LTE, 3G or 2G, the access link needs to carry these additional traffic flows also. Usually, the capacity from these sources is much less, as for LTE sites 1 Gbps ports were mostly used, and for 3G and 2G even less capacity is needed.

Fronthaul dimensioning is different due to time and loss sensitivity.

4.1.1.1 TDD UL/DL Split
The 5G standard includes both FDD and TDD modes. TDD uses the same frequency resources for both directions and the division is done by having slots for DL and UL

separately (e.g. 4 downlink, 1 uplink). On average, traffic in mobile networks is downlink dominated, although individual sites may experience a high uplink load for use cases – for example, when users are uploading files to the cloud – so especially in backhaul, transport links are loaded more in the downlink direction.

4.1.1.2 Fronthaul

On backhaul and midhaul, beamforming and MIMO functions are all terminated before being transmitted over the transport interface (F1 or N3). On fronthaul, the situation is different as the capacity depends on the functional split deployed and on which bits are actually transmitted over the low-layer split point. Fronthaul capacity needs additional information on the configuration and functional split between the Distributed Unit (DU) and Radio Unit (RU) that is not needed for backhaul or midhaul.

Another difference is that the scheduling function in the baseband, and the retransmission functionality with Hybrid Automatic Retransmit reQuest (HARQ), both introduce requirements for low latency of the fronthaul. Downlink and uplink capacity demands are also closer to each other than in backhaul, depending on the functional split and implementation.

4.1.2 Backhaul and Midhaul Dimensioning Strategies

In a new system, initially the amount of traffic is low before the new terminal penetration grows and transport links are also lightly loaded. Gradually, traffic picks up and a faster network – and faster UEs – invite more usage. Once transport links and nodes are deployed and connectivity services established, changes like capacity additions require extra effort, possibly site visits and even new installations. So, it is required to have a projection for growth with backhaul and fronthaul.

Theoretically, the transport link capacity is found by using a target input traffic mix and then calculating the resulting transport capacity with a dimensioning tool. The difficulty here is that the traffic volume and service mix may not be available as inputs, in which case other methods are used.

An alternative is to match the transport capacity to the air interface capacity. At the extreme, transport links are provisioned to peak air capability, always being able to supply the maximum multi-user MIMO throughput.

In practice, backhaul and midhaul transport links are often dimensioned according to realistic traffic needs, with some margin for growth to avoid frequent upgrade cycles. The target capacity value is compared against Ethernet line rates that are available for the site at a reasonable cost. Supplying transport capacity up to the theoretical peak air interface capability at every site leads to under-utilized transport links, especially in the early phase when there is still low traffic. This is avoided by taking an intermediate value between the sector average and sector peak as basis. Monitoring the traffic load becomes important, as transport link bottlenecks can then be discovered.

Dimensioning the transport link in this manner requires detailed site (radio) configuration, estimation of the protocol overheads and selection of the dimensioning rule which takes peak and average sector capacity into account. Example dimensioning rules for a single site are shown in Figure 4.2.

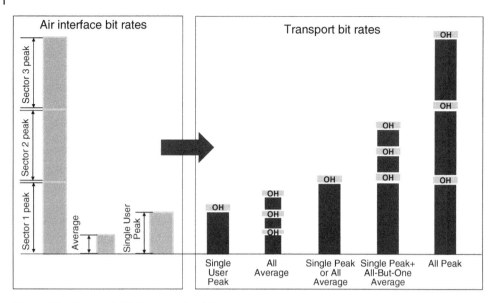

Figure 4.2 Transport 5G dimensioning [1].

The single-user peak depends, in addition to the sector conditions, on the UE capabilities also (e.g. how many transmit and receive antennas are supported). UEs evolve and then accordingly the single-user peak rate grows. The single-user peak rate is often used in marketing, and then it serves as a minimum that the transport link should be able to support in all cases.

With 5G, multi-user MIMO capability is assumed with sector peak achieved when multiple users get the single-user peak rate, until the air interface resources are consumed. The sector average rate is assumed to be some ratio of the peak (e.g. 0.1, 0.3). Supporting the higher value resulting from either one peak load sector or the sum of all sectors on average often leads to a reasonable value and is used in the following example.

4.1.2.1 Transport Media

Most commonly, Ethernet interfaces are used with fibre or wireless media supplying the same-capacity uplink and downlink. When the capacity in the transport link is the same in both directions and the air interface traffic volume is downlink dominated, downlink is the bottleneck direction in the transport network.

For transport media where the capacities differ in downlink and uplink, the above may not hold. An example is PON (Passive Optical Network), where uplink typically has less capacity available.

4.1.3 Protocol Overheads

Protocol overheads are specific to the logical interface in question and also depend on the transport configuration (e.g. whether IPsec is used and whether it is the IPv6 protocol or IPv4 on backhaul and midhaul). As the user plane consumes the vast majority of the capacity, the user plane overhead is essential. On top of the estimate for user plane traffic volume, an allocation for other traffic types – like control, management and synchronization plane – is added.

Sometimes, for transport links, the protocol overhead is calculated only taking into account the transport layers; sometimes it also includes higher-layer (radio) protocols. Transport includes IP, IPsec and Ethernet layers. Other protocols, depending on the logical interface, include GTP-U and UDP, and then SDAP/PDCP for the 5G RAN internal F1 interface. For dimensioning a logical interface for transport link capacity, the full protocol stack overhead needed to carry the user (UE) payload should be accounted for, including 5G radio-layer protocols and the tunnelling protocol.

The overhead due to protocol headers is often expressed as a percentage of the payload that is carried. With small user packets the overhead percentage is large and with large packets the percentage is smaller. An assumption for the packet size mix is needed. Assuming downlink is the bottleneck direction, it is expected that in peak-load downlink scenarios most packets are large packets. The overhead is added as an average based on the expected packet size mix.

Protocol headers (GTP-U/UDP/IPv6/Eth) sum up to about 100 octets. With IPsec tunnel mode, the overhead grows to around 180 octets, depending on the type of algorithms used. A further overhead is introduced in case fragmentation is required, as then additional header fields are needed.

Without fragmentation, for a 1200 octet payload the overheads in the downlink direction amount to ~8% without IPsec (IPv6) and ~15% with IPsec (IPv6). Not all packets are large packets, even in a peak-load situation, and fragmentation may still be needed in some links, so a 15–20% overhead could be assumed.

In the F1 interface in addition there are SDAP and PDCP protocol layers carried. The headers amount to ~4 octets, which is insignificant compared to the GTP-U/UDP/IP/Eth overhead; the F1 and NG overheads are approximately the same.

4.1.3.1 Other Traffic Types in the Link

When the user plane capacity demand including overheads is calculated, control and management plane data flows are added. The volume of control plane flows depends on the service mix and mobility assumptions. Analysis is provided for LTE in Ref. [2]. The control plane adds up to some percentage points of the user plane traffic.

The traffic volume for the management plane can temporarily be high (e.g. during SW downloads). Even at peak traffic load moments, some capacity of the management plane should be available so as to be able to perform remote management.

Synchronization with the PTP protocol requires a very small traffic volume and is in practice negligible, but requires high-priority treatment in the network.

4.1.4 Backhaul and Midhaul Capacity

Figure 4.3 shows an estimate for NG and F1 bit rates based on the air interface, showing sector peak rates and single-user peak rates and then two dimensioning results: Alternative A based on the higher figure of one sector peak or all sectors average and Alternative B based on single-user peak rate, which serves as the floor.

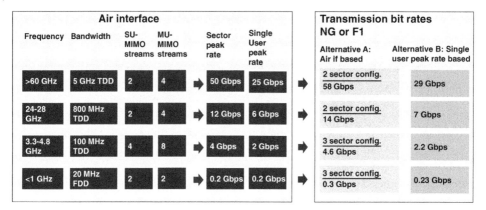

Figure 4.3 NG and F1 bit rates.

For FR1 and <1 GHz bands, a three-sector configuration is assumed and for FR2 and higher bands, a two-sector configuration.

For FR1 deployment with 100 MHz bandwidth and three-sector configuration, the capacity requirement is 2.2–4.6 Gbps for either backhaul or midhaul, depending on the dimensioning strategy selected, leading to 10 Gbps being the assumed line rate. In LTE, the most common rate was 1 Gbps.

For FR2 deployment with a two-sector example, the capacity needed amounts to 7–14 Gbps.

4.1.5 Fronthaul Capacity

Fronthaul transport delivers low-layer data bits between two 5G radio network elements, DU and RU. Part of the radio-layer computing is handled by the RU and part by the baseband.

The capacity requirement in fronthaul is higher than for midhaul or backhaul, since now the data transferred is not based on user packets but rather data is a partly L1 processed bit stream intended for the RU which completes the L1 processing. This L1 bit stream occupies more bandwidth than the user IP packet encapsulated in GTP-U. It also has additional requirements for delay, since air interface scheduling takes place in the baseband which is 'behind' the fronthaul and the HARQ loop operates over the fronthaul.

Functional split details are defined in O-RAN [3]. O-RAN option 7-2 is a frequency domain IQ sample stream. Other alternatives and sub-options are possible, leading to a variation of the transport link capacities. In the downlink direction, with Category A RUs precoding is in the baseband, while with Category B RUs the precoding is supported in the RU. In the uplink direction, split option 7-3 moves IRC (Interference Rejection Combining) to the RU. These all impact fronthaul transport capacities.

In general, transport capacity is reduced when more functionality is embedded within the RU. The main downside of adding functionality to the RU is that then the RU becomes a more complex network element and as a consequence there is less computing that can be centralized to a baseband hotel or to a radio edge cloud. Higher bandwidth (100 MHz in FR1, 800 MHz in FR2) increases the fronthaul bit rate. Increasing the MIMO layers does that as well.

FR1 100 MHz		FR2 800 MHz
8 streams	16 streams	2 streams
13-20 Gbps	25-40 Gbps	25-40 Gbps

Figure 4.4 Fronthaul capacity for FR1 100 MHz bandwidth.

Functions performed by baseband and RU in both downlink and uplink per split point were introduced in Chapter 2.

The transport capacities are shown in Figure 4.4 for a 100 MHz air interface bandwidth in FR1 and a 800 MHz air interface bandwidth in FR2.

With Category B RUs the transport capacity requirement is reduced from the above to approximately half, assuming that the number of layers is half the number of antenna ports.

For transport networks a strict real-time service is required as the scheduling occurs in the baseband. This means that queuing in fronthaul with large buffers is not feasible, and it also limits network-level statistical multiplexing gains unless it is known that sites do not experience peak demand at the same instant, which generally cannot be assumed.

In fronthaul, delayed packets beyond the arrival window are useless. If the scheduled slot in the air interface has already gone when the packet arrives from the fronthaul, the packet is discarded, so exceeding the fronthaul delay bound translates to fronthaul packet loss. Packet loss triggers retransmission between the UE and the baseband. Packet loss, fronthaul congestion and excessive delay cause air interface performance degradation as the air interface resources are consumed to compensate for fronthaul imperfections. This drives fronthaul towards a design where both delay and loss can be tightly controlled. The related profile is defined in IEEE802.1CM [4].

The fronthaul bandwidth requirement consists of a user plane sample stream and additional fast control data embedded within the user plane traffic flow. The other flows are for network management and synchronization (in case of packet-based timing).

4.1.6 Ethernet Link Speeds

In practice, Ethernet interfaces and Ethernet links are used both in backhaul and in fronthaul, with standard Ethernet line rates that are identical for both directions.

With multiple input sources, transport links cannot be dimensioned up to a full nominal link rate as the delay grows significantly in the lowest-priority queues when the link utilization exceeds 70–80%. Taking this into account, the Ethernet line rate is selected to meet the current requirements and projected growth for some time ahead.

Figure 4.5 depicts example transport capacities against different Ethernet line rates.

For backhaul and midhaul (F1 and NG2), an FR1 100 MHz three-sector gNB configuration can use a 10G Ethernet port. So while the LTE initially used mostly 1G line rates for backhaul, in 5G the comparable line rate is 10G.

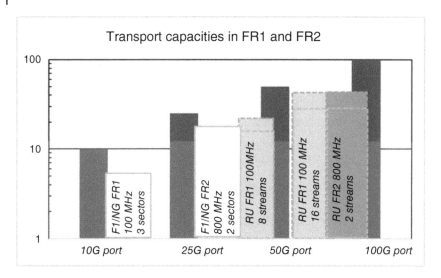

Figure 4.5 Capacity estimate for FR1 and FR2 and common Ethernet port rates.

In FR2, with an 800 MHz two-sector configuration, the capacity requirement grows over 10 Gbps, suggesting a 25G port.

For fronthaul, capacities are higher and 25G and 50G ports are used.

Multiple Ethernet ports, like 10 Gbps rates, can be grouped for higher capacity with the help of link aggregation. Ethernet link aggregation [5] allows binding member links to a link aggregation group, which then appears as a single interface.

The Ethernet is known for its continuous evolution towards higher line rates. The rates are achieved by first increasing the speed of the single-lane rate and then using those lanes in multiples, e.g. 2×, 4×, 8×.

After 25 Gbps, the single-lane rate for 50 Gbps is standardized as IEEE 802.3cd, using PAM-4 modulation [6]. The 50G line rate is the first with modulation instead of NRZ (Non-Return Zero).

50G lines could be used for 100G or 200G. The standard for single-lane 100 Gbps is ongoing by IEEE 802.3ck [7], allowing 200 Gbps or 400 Gbps. 1.6T could be a further step.

The other required characteristic of Ethernet is its low cost. Typically, the highest-rate ports are initially very expensive until the volume picks up and the pricing moderates, so selection of the line rate for backhaul and fronthaul is in practice impacted by which rates are economically feasible at the time of deployment. Then, in higher network tiers, multiple 10G links from the access require higher-capacity uplinks, like 100G.

In many cases the access site is shared and multiple radio technology base stations are served by the same uplink, with site switch consolidating all traffic flows. These other technologies may introduce typically up to some few gigabits per second of additional flows.

Solutions like PONs and microwave/wireless (including Integrated Access and Backhaul [IAB]) may offer service rates that do not directly match those of Ethernet line rates. The client device, a base station, may also interface a service provider side device over the UNI, which supports an Ethernet service less than the line rate (e.g. an Ethernet line rate of

25 Gbps is used for a 15 Gbps Ethernet service). The client-side egress interface is shaped to the correct service rate, to avoid losing packets due to buffer overflows or policing action on the provider side.

4.2 Transport Delay

4.2.1 Contributors to Delay in 5G System

One of the targets of 5G is reduction in latency. Latency reduction concerns user plane data flow but also context setup latency in the control plane.

URLLC introduces further requirements for low-latency services. Transport adds a significant delay component to all traffic flows simply due to the signal propagation time of 5 µs/km in fibre links. For wireless transport links, the propagation time is 3.4 µs/km so from a propagation delay point of view wireless has an advantage.

On the end-to-end service delivery chain, many components cause delay:

- Protocol processing in UE, gNB and core network
- Application server processing
- 5G air interface
- 5G logical interfaces (transport propagation delay and transport node delays)

Transport propagation delay consists of signal propagation delay that is specific to the media. Node delay consists of delay inside a node, including:

- Ingress and egress protocol processing (IP, Ethernet, IPsec, etc.) time
- Egress scheduling and queueing delay, specific to the Quality of Service (QoS) class
- Serialization delay
- Propagation delay

Transport delay is tailored by appropriate QoS marking so that intermediate nodes can schedule the flow accordingly with the selected per-hop behaviour. This optimizes delays in scheduling and queuing.

Additionally, a path through the network can be selected by traffic engineering to optimize delay and also a suitable local mobile network node (UPF) can be selected to allow shorter propagation delay for URLLC services.

The same transport network can thus serve multiple use cases and network slices which have different requirements for latency.

4.2.2 Allowable Transport Delay

The target value for transport latency is a question of first allocating the total end-to-end system delay budget to the contributing components. The end-to-end delay is a performance metric impacting network implementation and also is slice and service specific. The transport network, to the extent it is common for all services and slices, like it often is, has to meet the requirements for all services and all slices.

Fronthaul has additional requirements due to the HARQ protocol [8] operation between the UE and the baseband and air interface scheduling. Even if the end-to-end system budget would allow longer latency, radio protocols operating over the fronthaul mandate another, typically more stringent latency requirement for the fronthaul.

Figure 4.6 illustrates where the delay requirements originate and how the transport delay target values can be reached.

Requirements are split into end-to-end requirements and RAN protocol requirements. An example allocation of e2e delay budget to constituent components was presented in Chapter 2, with Table 2.3 having values allocated for targets of e2e Round-Trip Time (RTT) of 4.4 ms and 1 ms. These targets mandate a very short transport delay and local computing.

The end-to-end requirements primarily drive backhaul and midhaul interface targets. They are also valid for the fronthaul as it is in the end-to-end path, but in most cases the fronthaul itself already needs very low latency and the end-to-end requirements on fronthaul are met when the fronthaul meets the requirements from radio protocols. Time-critical radio functionality is in the baseband (DU) and that is why RAN protocols only impact fronthaul and not backhaul or midhaul.

Additional considerations originate from control, synchronization and management planes for tolerable delay. Often the user plane has the most stringent requirement for delay, but that may not always be the case.

Control plane latency is important as the delay incurred in the control plane signal path leads to context setup delays that are annoying to users. Transport delay on the control plane path must be low enough to also meet these setup-time requirements. The context setup consists of messages with multiple round trips.

Synchronization, when supported by packet timing (PTP), requires special consideration so that the network path for PTP supports low enough packet delay, delay variation and especially delay assymetry between the uplink and downlink directions.

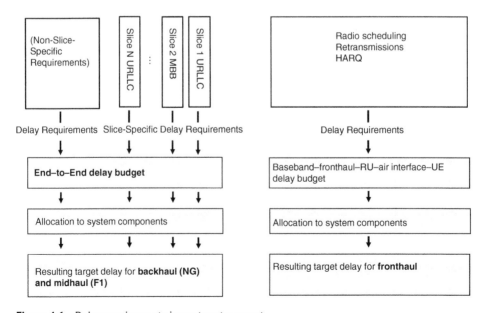

Figure 4.6 Delay requirements impact on transport.

When the transport network is in normal operation with no failures, transport delays are mostly constant except for variance due to buffering in the network nodes. During link or node failures, connectivity is disrupted and the topology may change after recovery, with a new transport path having a new delay value. This new value should still meet the delay requirement, which is a concern in fronthaul, where the delay budget is tight. Delay on the new path still has to be within the fronthaul delay bound.

In the access tier, typically all traffic types use the same first-mile access link from the cell site and all traffic types experience almost the same delay, which varies only due to different traffic classes. Over the end-to-end path for a specific network slice, network element peers can be located at different sites so propagation delay end-to-end is different for each peer – like UPF, AMF or network management. As an example, the user plane function of the Central Unit (CU) (CU-UP) may be located closer to the cell site while the control plane function of the CU (CU-CP) can be further, with the E1 interface in between.

With 5G there is more flexiblity in location of the peer entities as both user and control plane functions may be located either at a central core site or closer to the cell site, edge or far edge, and different network slices need their own connectivity. Cloud networking and virtualization make the location of the computing functions even more dynamic, and this also affects the delay performance both for the network slice and for the non-sliced cases.

4.2.3 User Plane and Control Plane Latency for the Logical Interfaces

For the user plane, the signal path in the SA mode is NG3 or NG3 + F1, depending on the architecture. In the NSA mode 3x the signal path is for split bearers S1 (or S1 + F1) for the 5G leg and S1 + Xn for the LTE leg.

In the control plane, the signal path in SA mode is NG or NG + F1, depending on the architecture. In the NSA mode, the control plane signal path for split bearers is S1 + Xn, or S1 + Xn + F1 for the 5G leg and S1 for the LTE leg.

Example values as targets for RTT are given in Table 4.1, taking into account mobile broadband use case with transport targets allowing wide-area network deployment.

The target value example allows transport interfaces to contribute to around 8 ms of the total e2e budget and support a distance from the cell site to the core of ~800 km, not taking into account any processing delays and with no allocation for real cable route distance, which could be 1.3 or 1.4 times line of sight.

This target is non-slice specific. Network slicing (e.g. for URLLC) introduces slice-specific delay target values.

Table 4.1 Example RTT target values (ms).

SA mode		NSA mode 3x (5G leg)		NSA mode 3x (LTE leg)	
Classical	CU–DU split	Classical	CU–DU split	Classical	CU–DU split
NG: 4–8	NG + F1: 4–8	NG: 4–8	S1 + F1: 4–8	S1 + Xn: 4–8	S1 + Xn: 4–8
	NG: 2–4		S1: 2–4	S1: 2–4	S1: 2–4
	F1: 2–4		F1: 2–4	Xn: 2–4	F1: 2–4

The distance from the core network and the distance from there to the server is a key contributor to the RTT, and bringing the core network function as well as the server closer to the cell site reduces latencies. With the core network function at the cell site (local core), transmission latencies of the NG interface are eliminated. This helps the URLLC slice case.

End-to-end RTT targets are not hard requirements but performance objectives. Also, transport networks on backhaul and midhaul tolerate longer delays, which then increase the e2e latency experienced. Values are implementation dependent.

Control plane latency impacts call setup times and reduced control plane latency is one of the 5G targets, as documented by ITU-R report M.2410 [9]. For the control plane a value of 20 ms is documented, with a suggestion for 10 ms.

In the CU–DU split architecture, the RRC messages are conveyed over the F1 interface. This means that in order to reach 10 m setup time, the F1 logical interface should not consume more than 2–3 ms one way.

4.2.4 Fronthaul (Low-Layer Split Point)

Low-layer fronthaul is different from the midhaul and backhaul interfaces, due to the real-time functions located partly in the baseband and partly in the RU, making the low-layer splint point (fronthaul) in between a strict real-time interface.

This requirement originates from air interface scheduling and HARQ processes located in the DU. Fast transmissions over the air interface take place over the low-layer split point, assuming O-RAN split point 7-2x.

As in Figure 4.7, the HARQ retransmission loop covers not only the 5G air interface but also the links and nodes on the fronthaul, like fibre links, optical active or passive elements and possibly switches and routers.

Issues impacting the maximum delay for the low-layer split point include the following.

- Transmission Time Interval (TTI): In the FR1 the TTI is 0.5 ms and in the FR2 it is 0.125 ms, as in Figure 4.7. The shorter the TTI the shorter the time between air interface transmissions and transmission on fronthaul.

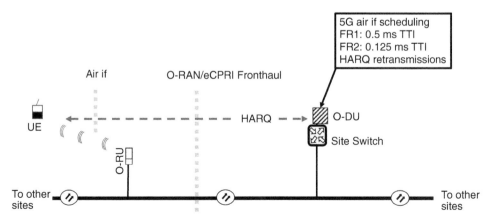

Figure 4.7 Real-time operations over the low-layer split point.

- Number of parallel HARQ processes: The more there are multiple HARQ processes running in parallel in the UE, the more delay can be tolerated without loss of throughput. As an example, there are eight HARQ processes in FR2 and four in FR1. This depends on the UE.
- Processing delay in UE and baseband: Some amount of time must be given to processing in the UE and baseband, which then reduces the delay budget to other contributors like fronthaul.
- DL/UL ratio (TDD): Different DL/UL ratios, such as 9:1, 8:1, 8:2. A higher DL/UL ratio improves the delay tolerance.

In IEEE 802.1CM a maximum latency of 100 μs (one way) is defined for fronthaul, where the target is to support full performance [4]. Depending on the factors above and on implementation, a few hundred microseconds of one-way delay may be tolerated.

As signal propagation on fibre is approximately 5 μs/km, this requirement limits the distance between O-RU and O-DU to some tens of kilometres, each 10 km physical distance consuming 100 μs for the signal (RTT) propagation alone and additionally some buffering or processing delay has to be allowed on the path depending on the network implementation. So, not all of the latency can be allocated to the signal propagation delay.

Converting the allowed latency to the maximum distance, the true length of the optical cable has to be considered as optical cabling seldom follows line of sight.

4.2.5 Low-Latency Use Cases

5G use cases demanding low latency in a few millisecond range cannot be supported over long physical distances simply because of the propagation delay on the transport links. To reduce latency, the whole 5G network as well as the application server as the communication peer to the UE is brought to a site close enough to the RU and the UEs, leading to a collapsed network architecture.

Different latency variations are presented in Figure 4.8. When the RTT of 1 ms is targeted, RAN, core and application server are on-site. With a few milliseconds target, the far edge site is used and for up to 10 ms the edge cloud site. If the RTT can be more than 10 ms, the core cloud can be used.

5G RAN functions may be disaggregated for CU, DU and RU, or some of the network elements may be collapsed to a single one. The transport links needed depend on the selected deployment type:

- If all functions are on the same site (RAN and core), the interfaces are site-internal and there is no need for an external transport link.
- O-RU remotely with O-DU and CU co-located: O-RAN eCPRI interface used.
- O-RU co-located with O-DU: F1 interface used.
- O-RU site, O-DU site and CU site all separately: fronthaul and F1 interface.
- Whole gNB functions at the remote site but core separately: NG interface used.

The flexibility creates different sets of interfaces and transport links for the design. The transport delay budget is allocated to each of the interfaces that are used in the selected architecture alternative.

With on-site or far-edge deployments, the geographical area is necessarily limited and also the whole low-latency transport network is of limited area – a plant floor, campus area

RTT Target

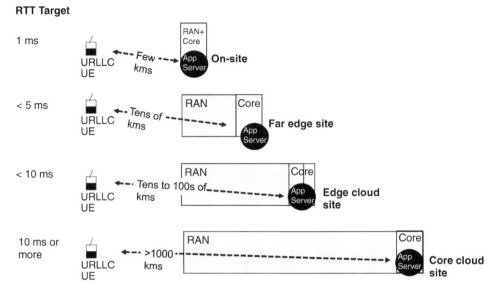

Figure 4.8 Deployment with local UPF.

or similar. This shortens distances on the transport links, and a separate local area back-haul or fronthaul network may be feasible. Otherwise, it is expected that the same transport network serves all use cases.

4.3 Transport Bit Errors and Packet Loss

4.3.1 Radio-Layer Performance and Retransmissions

In a mobile network, 5G included, the main obstacle for successful packet delivery is the air interface which is lossy, with high bit error and block error rates. These are addressed among other functions by radio layer retransmissions, with retransmitting protocols like HARQ and RLC (in acknowledged mode) [10].

In the networking domain there are no retransmissions for lost user plane packets. For the control plane, a reliable protocol (SCTP) is used as in the LTE, supporting retransmissions.

A summary of retransmissions is presented in Figure 4.9.

HARQ is responsible for compensating poor air interface conditions with retransmissions. The HARQ loop is between the UE and the O-DU (baseband), so it includes the low-layer split point. Another protocol, RLC, also supports retransmissions between the UE and the RLC entity in the baseband.

Packet loss on F1 or NG transport links has to be managed by the application layer between the user and the peer. If the application relies on a reliable protocol such as TCP or QUIC, there will be retransmissions (or lost data sent in new packets) between the UE and the peer through the whole 5G system. Packet loss on F1 or NG reduces the throughput to the UE with behaviour depending on the algorithm variant used by the reliable protocol.

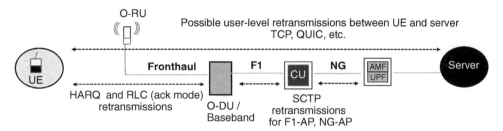

Figure 4.9 Retransmissions.

As seen in Figure 4.9, the low-layer split point (fronthaul transport) is included in the retransmission loop for both HARQ and RLC. In case a packet is lost on the fronthaul, this will be compensated by the air interface retransmission. This consumes the resources of the air interface as well.

For the control plane, SCTP is a reliable protocol, and it is defined for F1 and NG2/S1. With prioritization of the control plane packet flow in transport networks (e.g. AF class), the control plane should not lose packets even with congestion.

4.3.2 Transport Bit Errors and Packet Loss

Compared to the air interface which may have a packet loss ratio of e.g. 1..10% before retransmissions, a well-designed transmission link based on fibre or wireless operates with a residual bit error rate of 10^{-12} or less.

This low bit error rate in transport links may be achieved directly or with the help of error-correcting methods such as FEC (Forward Error Correction). FEC is included in many IEEE Ethernet standards for optical media.

Transport link bit errors lead to packet loss as erroneous frames are discarded at the link layer due to Ethernet L2 checksum failures before frames get forwarded to higher layers.

Packet loss occurs additionally due to congestion in network nodes, switches and routers, and is first experienced by packets in the lowest-priority queues. In the low-layer split point with time constraints, valid packets will also be discarded if they arrive out of the receiving window as after the packet is scheduled by the baseband, it needs to be received in time at the RU for transmission over the air interface and likewise for the uplink. In low-layer fronthaul, delayed transport packets turn into packet loss, which is then compensated by retransmissions by radio protocols (mainly HARQ, but also RLC).

Fronthaul bit errors induce packet errors. Assuming no error correction, a single bit error corrupts the whole packet and leads to the packet being discarded. With B_e as the bit error rate, P_e as the packet loss rate and N as the size of the packet, we have P_e as shown in Eq. (4.1), assuming bit errors are noncorrelating. Bit errors correlate in many failure cases, however residual bit errors can be assumed to be random.

$$P_e = 1 - (1 - B_e)^N \tag{4.1}$$

Table 4.2 Bit error and packet loss rate.

Interface	BER[1]	Packet loss rate[2]
Fronthaul (low-layer split point)	Less than 10^{-12}	Less than 10^{-7}
F1 (midhaul) – U plane	Less than 10^{-12}	Less than 10^{-7}
NG, S1 (backhaul) – U plane	Less than 10^{-12}	Less than 10^{-7}

1) Ultra reliability may impose stricter requirements.
2) Ultra reliability may impose stricter requirements.

For example, for a 1500-byte packet with $B_e = 10^{-12}$, $P_e = 1.2\times10^{-8}$. With smaller packet sizes, the contribution from the bit error rate becomes smaller. This eases the case for URLLC if those packets are of small size (e.g. short instructions to machines or actuators). For a 150-byte packet, $P_e = 1.2\times10^{-9}$ with the bit error rate assumed to be the same.

The bit error-induced packet loss rate is the floor that can be reached with zero congestion. For nondeterministic service, some amount of packet loss may exist due to congestion and resulting buffer overflows or also due to exceeding delay bounds. This is why time-sensitive networking targets minimizing congestion and controlling latency, for time-critical traffic streams.

The target for the packet loss rate depends on the use case and service. With mobile broadband the impact is on the end-user experience. With ultra reliability, the targets for end-to-end packet delivery via the 5G system are much more stringent, depending on the application.

4.3.2.1 Transport Links/Summary

Bit errors and packet loss on transport links should have a minimal impact on the end user. The bit error rate should be low enough to allow ultra-reliable service through the link. Packet loss should also be low enough for the URLLC and also low for the MBB service, so as not to impact air interface and user experience.

Typically, the same network links support all 5G use cases, so URLLC needs to be considered. Apart from the local deployments discussed, it is not economical to consider separate transport networks for high-performance URLLC use cases.

In the Table 4.2 example, the bit error rate $<10^{-12}$ allows some room for packet loss due to congestion (buffer overflow or discards due to delayed packets), as the packet loss target is $<10^{-7}$.

4.4 Availability and Reliability

4.4.1 Definitions

4.4.1.1 Availability

3GPP defines the communication service availability as the 'percentage value of the amount of time the end-to-end communication service is delivered according to an agreed QoS, divided by the amount of time the system is expected to deliver the end-to-end service according to the specification in a specific area' [11].

For availability A, the formula is described in Eq. (4.2):

$$A = \frac{\text{Operational time}}{\left(\text{Operational time} + \text{downtime}\right)} = \frac{\text{MTBF}}{\left(\text{MTBF} + \text{MTTR}\right)} \tag{4.2}$$

MTBF = Mean Time Between Failures and MTTR = Mean Time To Repair. MTBF is an expected value based on some assumed distribution of failures and is no guarantee that an individual system works without failures for the duration of the MTBF.

Sometimes five nines is denoted as high availability, but it is not a formal definition.

4.4.1.2 Reliability

3GPP defines communication service reliability as the 'ability of the communication service to perform as required for a given time interval, under given conditions' [12].

Reliability is often calculated assuming exponential distribution with failure rate λ for a certain mission time t, Eq. (4.3). The exponential distribution is memoryless, meaning that when the system is in operation, it is expected to perform correctly just as long as a new system, no matter how long it has been in operation already. This assumption is valid when the system has passed the infant mortality phase and is in the plateau phase of the bathtub curve of failure rates, before the failure rates again increase at the end of life with wear and tear:

$$R = e^{-\lambda t} \tag{4.3}$$

This formula estimates the reliability of a component during time interval t when the failure rate λ is known. The failure rate is the inverse of the MTBF, Eq. (4.4):

$$\lambda = \frac{1}{\text{MTBF}} \tag{4.4}$$

The difference between availability and reliability is that reliability helps estimate the probability of success of a mission during a certain mission time. Availability also covers what happens after the failure, how long it takes before the system is back in operation again.

Further information can be found in Refs [13] and [14].

4.4.2 Availability Targets

Ultimately, what is interesting is having an availability target for the 5G service and any URLLC reliability objectives for packet delivery and then setting related targets for backhaul and fronthaul.

At the mobile system level, an analysis of availability taking into account core network and radio network and network elements is given in Ref. [15] for the 2G and 3G network. The principles outlined there can be carried out for the 5G network too, to find the bottlenecks and critical single-point-of-failures, including those in the transport network.

The availability definition depends on when the 5G system is still considered operational: is it needed that all gNBs are up in the network or is it allowed that some of the gNBs are down. If part of the gNBs can be down, then also the access link to that gNB can be down if connectivity to the other gNBs is working. Since cell areas overlap, UEs may be served by another base station during outage of the failing site. Network performance may be degraded, as there is less traffic-handling capacity and possibly other limitations. This, however, stresses the importance of having other, nearby base stations in operation if one fails.

With FR2 (millimetre-wave) frequencies, cells are small. The use cases in FR2 are in adding capacity to the network (hotspots) or providing coverage to specific areas like street canyons. With hotspots, typically a macro layer (FR1 or LTE) exists that provides service if the FR2 cell is down. With the coverage case there may be no other cell available. What helps in FR2 is that beamforming allows beams to be targeted and by this means another cell may reach the same location.

4.4.3 Availability in Backhaul Networks

Typical availability targets currently for last-mile transport links are on the order of four to five nines. The mean down time per year would be 52.56 min for four nines. Backhaul network availability end-to-end, including all links and nodes in the signal path, is limited by the cell site access link (the last-mile link), which is rarely protected.

In higher tiers of the transport network resiliency is easier to arrange and also more essential due to aggregation of traffic from several cell sites. Hub sites in access and aggregation tiers are examples of critical sites, as well as any baseband hotel or far-edge sites. Loss of these sites or the site uplinks causes outage for an area and so resiliency is critical. A target minimum of five nines availability is usually needed.

At these higher tiers of the network, topology is however built for resiliency and alternate paths are available. The dominant technology is IP/MPLS over optical infrastructure and for example with IP VPNs, commonly alternate paths are supported at the outer tunnel level in the provider network based on IGPs (Interior Gateway Protocols). Attachment to the IP VPN service is also resilient using dual attachment, so single-point-of-failures can be avoided.

Causes for transport failures are cable cuts and other link failures, node failures, power outages but also human errors like misconfigurations.

Estimating failure rates is difficult. It can be based on data collected from actual networks. The accuracy of the calculations for availability or reliability depends on the validity of the assumptions. The methodology and formulas can be used by adjusting the input assumptions to match the case being analysed.

For fibre links, the failure rate can be estimated with a cable cut (CC) metric and the length of the cable (L), Eq. (4.5):

$$\text{MTBF} = \frac{1 \text{ year}}{L} * CC \tag{4.5}$$

CC gives the distance per one cable cut in a year. The MTBF decreases with increasing length of cable, as with longer distance it is more likely to experience a cut due to civil works, digging, etc. at some point along the cable route. See e.g. [14].

In Table 4.1 the availability of fibre cable of length 10 km is calculated using a CC of 600 km. The setup is shown in Figure 4.10, where case A and case B are compared; the latter with a duplicated link and automatic recovery, compared to the former with a single link and manual repair.

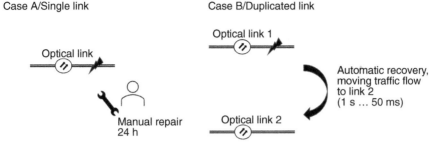

Figure 4.10 Availability of optical fibre link.

Table 4.3 Input assumptions and resulting availability estimate.

Case	MTTR	Availability	
A	24 h	99.995433998%	four and half nines
B1	1 s	99.999999947%	more than nine nines
B2	50 ms	99.999999997%	more than ten nines

In the first row of Table 4.3, 24 h manual repair time gives four and half nines, which might be adequate already for an access link for single gNB.

Moving to a protected fibre facility with recovery through protection switching or restoration with repair time MTTR of 1 s improves the availability to nine and half nines. Going to 50 ms recovery is possible with both optical and IP recovery, especially when the local recovery action is adequate to resume traffic, and 50 ms recovery increases the estimated availability to more than ten nines.

The example shows the improvement in availability resulting from moving from a single facility with manual repair operation to a protected facility with automatic recovery. Availability is already very high in the example, even with 1s recovery time.

The calculation covered only fibre. Optical interfaces, transceivers and possibly other nodes contribute to unavailability. These components could also be redundant to avoid single-point-of-failures.

For the result to be valid for case B, failures on the two optical links must be non-correlating, meaning that they must not depend on the same physical cable and share the same risks, but instead need to have diverse physical routes so that if the cable carrying optical fibres for link 1 fails, this does not imply or increase probability of failure for link 2. This requires knowledge of cabling and physical topology and planning the resiliency accordingly. The shared-risk link group concept is used to address this topic. The need to have independent cables increases the costs and is especially difficult in last-mile access links, where it may be challenging to get the fibre connection at all. An alternative is to use fibre complemented with another media, like wireless.

The big improvement comes when the manual repair time is avoided due to automatic recovery. Even with automatic recovery, manual repair action is still needed to correct the fibre link, but during that time the system is operational over the second fibre link and the traffic flow is not blocked. An added benefit of this is that since the outage is already resolved, there is less urgency to repair the failed cable, which can be a major cost saving. Other benefits of the reduced outage time are that there is less revenue lost directly and perhaps even more importantly, customer experience is not compromised.

Availability could in theory be increased by shortening the recovery time and then further almost arbitrarily by adding parallel, independent facilities. Limits come due to complexity and cost, as the network needs more links, nodes and sites, and more functionality too.

4.4.4 Recovery Times in Backhaul and Fronthaul

An important consideration in increasing availability is designing recovery target times from backhaul and fronthaul failures to be fast enough to have the least possible impact on 5G services. Examples impacts are:

- loss of user plane packet or few packets
- loss of control plane packet or few packets
- loss of ongoing call/context
- loss of whole logical interface

Previously, the case of packet losses and retransmissions was discussed. Single or a few packet losses may happen in transport due to congestion or also due to bit-error-induced packet discards. These are more performance degradations than full failures. Typically, transport failure is a link or node going completely down and in this case all packets are lost indiscriminately for the duration of the break.

A critical moment is the time after which the context (call) for UE is lost. This happens after a few seconds. For longer breaks of perhaps tens of seconds, 3GPP logical interface management procedures start to try to recover from the failure and reset the interface. Both values depend on implementation.

Telecom networks based on Sonet/SDH have set the standard for recovery time to 50 ms based on Sonet/SDH protection switching. This is achievable on packet networks as well, especially when a local decision is enough to mitigate the impact of failure (e.g. with fast reroute, Equal Cost Multipath [ECMP] or Ethernet link aggregation). Otherwise, recovery by routing protocols may take seconds, but can be sub-second too, depending on many factors, like routing protocol used, detection method, size of the network, etc. A lot depends on the failure detection time and the time needed to propagate the fault event to other nodes, and then recalculate and enforce the new active topology. When the target is to keep the ongoing context alive during transport recovery, breaks of some hundreds of milliseconds, or even a few seconds, are tolerated. This fits quite well with what is realistically achievable with networking recovery alternatives.

4.4.5 Transport Reliability

A new topic with 5G is ultra reliability and low-latency URLLC services. Services may need either ultra reliability or low latency, or both.

The reliability requirement is demanding when it is combined with the low-latency requirement, as there is then little time for corrective actions, like retransmissions or transport link recovery, when the initial packet delivery fails. Industrial applications may need an instruction packet (e.g. every 500 μs) and do not tolerate loss of more than a few successive packets, which means that recovery mechanisms on the transport network are not fast enough to correct the failure so that successive packets or a retransmission (if supported) would succeed in time.

One important use case for 5G is this type of industrial application. They vary a lot in requirements. As discussed in Chapter 2, a reliability target of $R = 1 - 10^{-5}$ was included in Release 15, and this was increased to $R = 1 - 10^{-6}$ in Release 16/17. Depending on the case, these packet delivery reliability levels could be achieved directly or with the help of packet duplication over multiple paths.

The difficulty of calculating availability or reliability lies in making assumptions that describe the situation correctly, and having a sufficient amount of data on which to base those assumptions. Each network and case is likely different, and assumptions for one case may not hold for another.

NGMN assumes the probability of failure to be 10^{-4} to 10^{-6} for individual links or nodes, based on a defined reliability target for delivering a packet from the Application Server (AS) to the UE [16].

For any transport link or node that is on the path of the URLLC packet delivery, the target reliability should be at minimum that of the air interface, preferably better so that transport does not compromise the system. This actually also leads to the range included by NGMN.

In order to have a view as to ultra-reliability requirements and transport, one may calculate an estimate for the reliability of a fronthaul fibre link between two sites: one is a collapsed site with application server and gnB baseband, and the other is a remote O-RU site, which is located at a 0.3 km distance as in Figure 4.11.

In Table 4.4, the CC metric is used to estimate reliability of the fibre and no other components like transceivers or transport elements on the path are included. Mission time is set to 10 years as the operating time of the system.

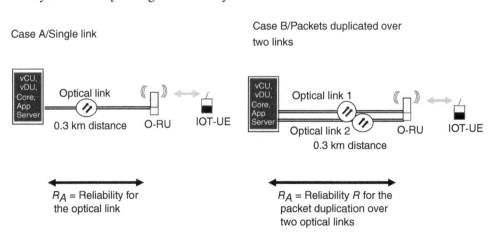

Figure 4.11   Transport reliability example.

Table 4.4 Input assumptions for example reliability calculation.

Cable cut metric (CC) =	600	km
Fibre link distance (d) =	0.3	km
Mission time t =	10	years

The basic case (case A) uses a single fibre link facility. In order to see the possible improvement using dual facilities with packet duplication, case B is calculated.

The failure rate is based on Eq. (4.6):

$$\lambda = \frac{1}{\text{MTBF}} = \frac{d}{\text{CC}} \tag{4.6}$$

For case A, the formula is based on Eq. (4.7):

$$R_A = e^{-\lambda t} = e^{\left(-\frac{d}{CC}*t\right)} = e^{\left(-\frac{0.3}{600}*10\right)} = 0.995012479 = 99.5\% \tag{4.7}$$

For case B, the formula is based on Eq. (4.8), where there are two independent parallel systems with identical reliabilities R1 and R2, so where $R1 = R2 = R_A$:

$$R_B = R = 1 - (1 - R1)*(1 - R2) = 1 - (1 - 0.995012479)^2 = 0.999975125 = 99.998\% \tag{4.8}$$

In case B, packets are copied over both links and the duplicate packets discarded at the receiver end. Case B assumes that the links do not share the same risks. Duplicating packets over two independent paths improves the reliability significantly, from 99.5% to 99.998%.

With packet duplication, the receiving end takes the first valid packet within the receiving time window and discards possibly later-arriving duplicates which have the same sequence number. Reliability is improved as now failure on one signal path does not cause failure of the mission. Multipath could be extended to more than two paths for even higher reliability.

While the case in Figure 4.10 relied on protection switching or other recovery that moved the traffic flow to another link when failure happened, in Figure 4.11 both links are used all the time and packets are duplicated. There is no need to move the traffic flow, as it is readily on two paths, and the receiving-end application experiences no packet loss during the failure, which happens in case of Figure 4.10 for the duration of recovery.

Reliability of the fibre link needs to be estimated based on the case; the use of a CC metric might not be a valid assumption. Further, it is worth noting that the analysis accounted only for the physical media (fibre) failures to mimic the approach in the air interface.

The result shows that multipath and packet duplication over two independent transport links significantly increases packet delivery reliability. It also shows that transport may be a critical component in the URLLC packet delivery path and, as such, transport cannot be omitted directly from the reliability estimates, as the resulting reliability for transport – even in the packet duplication scenario – was found to be less than $R = 1 - 10^{-5}$.

4.4.6 Air Interface Retransmissions and Transport Reliability

Air interface retransmissions generally help improve air interface reliability. In the URLLC environment, retransmissions have the difficulty that they necessarily take some time. The feasibility of the retransmissions depends on the latency requirement and type of air interface (TDD/FDD). With TDD, the slot structure means that acknowledgements or retransmissions have to wait until the next slot becomes available. This may be too long for a low-latency service.

In transport, there is no reliable service, as UDP/IP or Ethernet is used. However, HARQ and RLC (acknowledged mode) retransmission loops operate over the fronthaul (see Section 4.3.1).

Retransmissions help the air interface when radio channel conditions vary rapidly, so that a fast retransmission may succeed even if the original transmission failed.

The same is not true for typical fronthaul transport failures. If the link is down due to cable cut, retransmission within a time frame of milliseconds does not succeed unless the transport failure is corrected before the retransmission. The time available for transport recovery is very short, as even the 50 ms benchmark is two decades of magnitude too long. In special cases, hardware-based mechanisms against local failures may be fast enough, but generally transport recovery needs more time than is available in the HARQ retransmission cycle. This means that fast air interface retransmissions generally do not help in case there is a link break in fronthaul transport. Other solutions for increased transport reliability are needed.

For packet discards due to fronthaul transport link residual bit error rates or other random packet losses where single or a few packets get lost, retransmission likely succeeds if the anomaly on transport is a short one and goes away without repair action. Against that phenomenon, retransmissions with HARQ do help.

For the air interface with single retransmission, a simple model for reliability in packet delivery is based on Eq. (4.9), where p is the probability of a packet loss in the air interface for a single transmission.

$$R = 1 - p^2 \tag{4.9}$$

Adding multiple paths with similarly one retransmission increases reliability. Path diversity with two independent paths and one retransmission leads to reliability estimate based on Eq. (4.10):

$$R = 1 - p^4 \tag{4.10}$$

With the air interface, path diversity and retransmissions in combination increase reliability, in addition to many other 5G air interface improvements for URLLC.

The packet error rate for transport links due to random bit errors is very low, $<10^{-7}$ or even $<10^{-10}$ for small packets. Refer to Table 4.2 earlier. Critical traffic must also not experience congestion in the transport network, so the packet loss rate due to random bit errors and congestion summed together is $<10^{-7}$. The remaining issue with transport for the URLLC service is the risk of link breaks and node failures, and these will be discussed further with 5G system packet duplication.

4.4.7 Packet Duplication in 5G and Transport

The 5G system includes different multipath and packet duplication alternatives, some of which also cover transport legs. At the networking domain, technologies supporting packet replication services are also available as additional tools where required.

4.4.7.1 DCP Packet Duplication in RAN

Figure 4.12 shows duplication at the PDCP layer with dual connectivity, in case the CU and DU are co-located but the RUs are on remote sites. Two basebands (CU and DU) are on two different sites, as well as the two RUs on different sites, so there are also two different fronthaul links.

The PDCP packet is duplicated at the CU interfacing the core network (N3) and the duplicates are sent using two RLC entities via two paths, one via the local air interface and the copy via Xn to another air interface via the other CU/DU site. See Ref. [17] for the 3GPP definition of PDCP duplication.

Multipath with packet duplication now covers not only the air interface, but also the fronthaul. The peer PDCP entity is in the UE where the duplication is eliminated. For the transport network, this duplication functionality is not visible except due to the extra capacity consumed by the packet copy.

For reliability, the configuration in Figure 4.12 can be analysed with a reliability block diagram (Figure 4.13) in order to find the impact of the multipath and PDPCP duplication to packet delivery reliability, taking into account the transport contribution from the Xn and the fronthaul links.

The gNB sites are 0.3 km distance from each other and also the RU is 0.3 km distance from the baseband. Here it is assumed that the reliability for the 0.3-km fibre link is one decade better than was assumed earlier in the case of Figure 4.12, and so $R1 = R21 = R22 = 0.9995$. An air

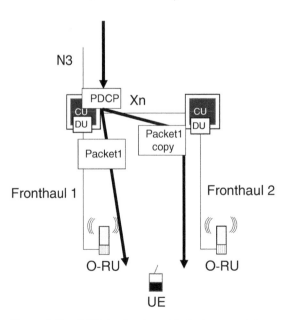

Figure 4.12 PDCP duplication with dual connectivity.

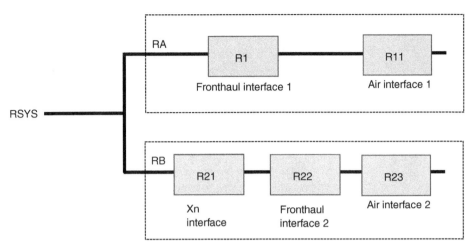

Figure 4.13 Reliability block diagram including Xn and fronthaul.

interface packet loss rate of 10^{-5} is assumed, and the target is to calculate an estimate for the reliability of packet delivery over the system.

The system consists of two subsystems, with reliabilities RA and RB, in parallel configuration. For the complete system, an estimate for the reliability is based on Eq. (4.11):

$$
\begin{aligned}
RSYS &= 1 - \big((1-RA)*(1-RB)\big) = 1 - \big((1-R1*R11)*(1-R21*R22*R23)\big) \\
&= 1 - \big((1-0.9995*0.99999)*(1-0.9995*0.9995*0.99999)\big) \quad \text{(4.11)} \\
&= 0.999999485 = 99.99995\%
\end{aligned}
$$

The system as a whole reaches very high reliability of more than $1 - 10^{-6}$ due to the packet duplication over two paths, even with the lowest component reliability, transport in this case, being only 0.9995.

In order for the transport links not to have an impact, the target for transport – including all links and transport nodes – should be about one decade better than that of the air interface performance, or 0,999999 in the example. The transport system could have this increased reliability as a single facility, or if that is not achievable, then by the help of a separate transport-level packet duplication service over disjoint paths.

5G RAN-level PDCP duplication nevertheless greatly increases the reliability as the path diversity covers both the air interface and the fronthaul. PDCP duplication can include up to four paths as defined in 3GPP, and four paths would increase the reliability further.

PDCP duplication with multipath connectivity can also be deployed in an architecture where there is no fronthaul but gNB is all integrated as a classical base station. This removes fronthaul from the reliability block diagram and the contribution it makes. Since each component in the chain introduces some probability of failure, reducing components in the packet delivery path increases reliability.

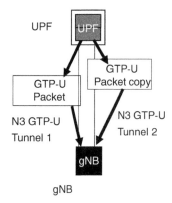

Figure 4.14 Redundant N3 tunnels with GTP-U replication.

4.4.7.2 Redundant N3 Tunnels

For the case where UPF is located remotely from the gNB, reliability on the N3 interface is relevant. An example deployment is where UPF and application server are located centrally, and then gNBs are remote in the campus area to provide coverage for the URLLC service.

As in Figure 4.14. a single PDU session can use two GTP-U tunnels that are routed in the transport network by different paths, as in the 3GPP specification [18]. Each tunnel uses a different IP endpoint and the network layer keeps the routes on different paths to avoid fate sharing, in order to realize the reliability gain. The packet is duplicated by the UPF and the duplicates are eliminated based on GTP-U sequence number at the gNB. GTP-U duplication again creates two parallel systems and related reliability improvement.

4.4.8 Transport Analysis Summary for Availability and Reliability

Availability and reliability have to be analysed starting from the service requirement and then breaking that end-to-end requirement for allocation to individual components, where transport is one, analogous to the process depicted in Figure 4.7 for the e2e delay requirement, related delay budget and allocation of delay budget to the contributing components.

For availability and reliability, 5G deployment type and system functionality (e.g. PDCP duplication, GTP-U duplication, dual connectivity and air interface retransmissions) all impact the analysis. The main difficulty is in finding valid input assumptions for estimating either availability or reliability for the transport components.

For transport, availability should target a minimum of five nines and preferably higher, depending on 5G service availability requirements. Availability is especially critical in cloud sites and hub sites and their attachment to the network in a resilient manner, and in links or nodes carrying traffic from multiple cell sites. Redundant paths and nodes with short, automatic recovery services increase availability.

For reliability, transport links normally operate with low enough residual bit error rates ($<10^{-12}$) in well-designed links. Packet losses due to congestion have to be eliminated for URLLC services so that packet losses due to congestion are also $<10^{-7}$.

Other sources for transport unreliability are link failures and node failures, which may impact packet delivery heavily. The 5G air interface has optimizations for the URLLC service, targeting very high reliability, and it is not necessarily trivial to match this on the transport side, at least when estimating it with input assumptions from wide-area links.

For URLLC services, packet replication over disjoint paths is a useful tool to mitigate these transport failures also – either at the 5G system level, like with PDCP or GTP-U, or separately at the networking layers. The target for transport reliability in case of URLLC services is at minimum that of the air interface, preferably higher so as not to compromise the system. Additional packet duplication services on transport links may be needed if the end-to-end target is otherwise estimated not to be met.

4.5   Security

User communications in the 5G network are secured with 3GPP-defined extensive security architecture, which covers – in addition to the user communications – also the logical interfaces between elements in the network domain.

The control plane from UE and the UE data flow are fully protected. Actually, there may be two or three layers of cryptographic protection for UE data flows, as NAS (Non-Access Stratum) signalling messages are carried transparently from the UE over the RAN to the core and protected, the underlying messaging in the RAN control plane segment is protected and below that in the network domain the logical interface is protected.

For transport and networking, protection defined by 3GPP for the network domain is essential.

4.5.1   Summary of 5G Cryptographic Protection

NAS signalling is protected end-to-end directly between the UE and the core network (AMF).

AS (Access Stratum) signalling is protected between the terminal and the radio network with the PDCP protocol located in gNB. PDCP is the protocol layer in RAN that includes encryption and integrity protection services.

Within the 5G RAN, the user plane from the UE is both encrypted and integrity protected (depending on configuration), which is an improvement over LTE, where the user plane is encrypted but not integrity protected. Between the RAN and core there is no protection in the user plane except for the network domain protection.

Due to the above, 5G logical interfaces carry flows that are already covered by 5G system security (such as NAS signalling) and flows that are not covered, like the user plane at the N3 interface. 3GPP-defined network domain security adds protection to the logical interfaces so that all flows are covered.

High-layer split point (F1): In the control plane from the UE AS and NAS signalling, both integrity and encryption services are supported. Control plane (F1-AP) traffic: DTLS and network domain protection are defined. For F1 user plane traffic both integrity protection and encryption services are supported.

(Xn): In the control plane from the UE AS and NAS signalling, both integrity and encryption protection are supported. For the Xn-AP DTLS and network domain protection is defined. User plane traffic: As with F1.

Backhaul (NG2): From the UE NAS signalling, both integrity and encryption protection are supported. For NG-AP, traffic DTLS and network domain protection are defined.

Backhaul (NG3): There is no protection as the PDCP protocol terminates at the access side of the gNB, except in the network domain.

Backhaul NSA mode 3x for the S1: From the UE AS and NAS signalling, both integrity and encryption protection are supported. For S1-AP traffic network domain protection is defined. User plane traffic: No protection as the PDCP protocol terminates at the access side of the gNB, except network domain protection is defined.

In NSA mode 3x, the PDCP layer on the gNB side is the anchor point which splits the traffic of the split bearer to the X2/LTE interface and to the 5G interface and the NR PDCP layer is used.

Network management flow is defined to be encrypted and integrity protected, with TLS and network domain security.

4.5.2 Network Domain Protection

For the network domain, the summary of Section 4.5.1 means that from the traffic flows:

- Radio network control protocols like F1-AP, NG-AP are not protected apart from the AS/NAS signalling messages from the UE, and need cryptographic protection within the network domain.
- The user plane on the backhaul interface (NG) is not protected since the PDCP termination is on the RAN side (as in LTE) and needs cryptographic protection in the network domain.
- User plane interfaces within the RAN (low-layer and high-layer split point) carry flows that are PDCP protected.
- Other traffic flows are unprotected unless addressed (synchronization, network management) in the network domain.

IPsec is defined as the technology in the network domain for security, as detailed in Ref. [19]. Both tunnel mode and transport mode are included, with transport mode being optional. IPsec can be integrated into a network element like gNB or security services can be provided by a separate element, a Security Gateway (SEG). Between security domains, an SEG is used.

4.5.3 Security in Fronthaul

In the low-layer split point (fronthaul), the UE AS and NAS signalling are both integrity and confidentiality protected. The user plane traffic of the UE is also integrity and confidentiality protected. The O-RAN fronthaul standard does not define additional network domain cryptographic protection.

4.6 Analysis for 5G Synchronization Requirement

With the introduction of the 5G system, the topic of synchronization must be carefully addressed, similar to the LTE system. In general, 5G RAN transmission requires synchronization to meet the following two error budgets at the Over The Air (OTA) interface:

- frequency error
- time alignment error

High-level analysis for the frequency error and time alignment error (for different features) are provided in the following subsections.

4.6.1 Frequency Error

Frequency accuracy is an important factor to allow the UE to continuously track the BaseStation (BS) reference clock signals in BS downlink transmission. The allowed frequency error is related to the UE PLL tracking capability (PLL and VCXO components) and the BS reference clock signal transmission (precisely the time distance between PSS and SSS transmission). *Note*: PSS = Primary Synchronization Signal, SSS = Secondary Synchronization Signal based on 3GPP NR standard TS 38.211 definition [20].

Table 4.5 5G system frequency error target.

BS class	Accuracy
Wide-area BS	±0.05 ppm
Medium-range BS	±0.1 ppm
Local-area BS	±0.1 ppm

Based on 3GPP NR standard TS 38.104 [21], the 5G system frequency error in Table 4.5 is applicable at the OTA interface. It is based on BS class and is independent of different features. It is also applicable to the FR1 range (<6 GHz spectrum range) and FR2 range (centimetre wave and millimetre wave spectrum range).

In general, with the fixed ppm accuracy specification (e.g. ±0.05 ppm), it is more of a challenge for the FR2 range due to higher carrier frequency. At 28 GHz carrier frequency, ±0.05 ppm corresponds to ±1.4 kHz frequency error. Without demanding better but more costly UE PLL tracking capability, the 3GPP NR standard TS 38.211 allows PSS/SSS to be sent with higher SCS configuration (120 kHz or even 240 kHz). This effectively helps UE PLL tracking to tolerate 1.4 kHz frequency error with good margin.

Note that the 3GPP NR standard decides to choose the same ±0.05 ppm specification on the wide-area BS class between 5G and LTE standard. This is ideal for LTE to 5G migration reasons.

For medium-range BS and local-area BS classes, higher frequency error is allowed due to lower end-user mobility in the corresponding deployment scenarios.

4.6.2 Time Alignment Error (Due to TDD Timing)

The 5G system can be based on FDD or TDD technology similar to the LTE system. Use of FDD or TDD technology depends on the available spectra in different world markets. For 5G target, the spectrum expansion into the centimetre wave and millimetre wave frequency range tends to expand the TDD technology deployment scenarios.

In TDD technology, downlink transmission (to UE) and uplink transmission (from UE) share the same frequency in the spectrum. Therefore, there is a TDD switching key concept to allow such sharing. At the high level, the 3GPP standard allocates a switching gap between downlink and uplink transmission to facilitate the switching. During the switching gap, BS and UE can turn on/off transmission and reception based on the agreed upcoming TDD event (transmission or reception for BS and UE). In addition, when there are neighbouring base stations in a geographical area, all base stations must switch together to avoid TDD interference. Time alignment among all base stations becomes an important requirement for TDD technology to work.

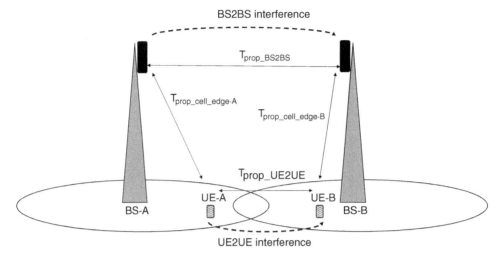

Figure 4.15 TDD interference aspects.

There are different types of TDD interference to consider. As shown in Figure 4.15, during each switching event (either DL to UL switching or UL to DL switching), between any neighbouring sites, there could be

- BS to BS interference
- UE to UE interference

5G synchronization time alignment error, BS and UE transmission switching time, cell radius (or ISD between sites) are all factors that affect the TDD interference. For each 5G SubCarrier Spacing (SCS) configuration, the achieved cell radius to avoid any type of TDD interference can be determined based on these factors.

An analysis is provided in this chapter to understand the impact of these factors. It is based on information from a 3GPP contribution to the RAN4 working group. Refer to R4-1703013 [22] for details.

The analysis uses the 3GPP hex model for cell definition to define neighbouring sites in Figure 4.16. The corresponding cell radius (Rs), InterSite Distance (ISD) and cell-edge propagation ($T_{prop\_cell\_edge}$) are defined.

Based on the TDD concept of sharing DL and UL transmission at the same frequency, there is always a switching gap (T_{GUARD}). The switching gap is further split into a DL–UL switching region and a UL–DL switching region by the relationship $T_{GUARD} = T_{DL\_UL} + TA_{offset}$. T_{GUARD} and TA_{offset} are system configuration parameters across all BS and UE. Based on the gap in these switching regions, there are four types of TDD interference to analyse.

(1) BS2BS interference in DL–UL switching (**BS2BS-DL2UL**), see Figure 4.17. Consider that BS-B timing is behind (i.e. later than) BS-A timing, then BS-B (aggressor) late DL transmission can interfere with BS-A (victim) start of UL reception.

To avoid BS-B end of DL transmission (point X) overlapping with BS-A start of UL transmission (point Y), we have the constraint in Eq. (4.12):

$$(T_{GUARD} - TA_{offset}) > T_{sync} + T_{BS\ on \rightarrow off} + T_{prop\_BS2BS} \tag{4.12}$$

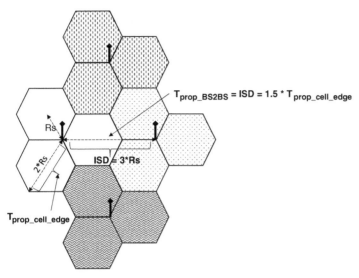

Rs = Cell radius
ISD = Inter-Site Distance
$T_{prop\_cell\_edge}$ = Propagation delay to cell edge
T_{prop_BS2BS} = Propagation delay from BaseStation to BaseStation

Figure 4.16 Hex model for cell radius, cell-edge propagation and intersite distance definition.

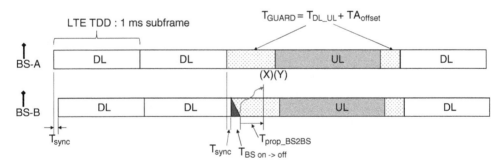

T_{GUARD} = TDD switching gap
T_{sync} = difference in BS-A and BS-B synchronization alignment
$T_{BS\ on\text{-}>\ off}$ = transition time to complete turning off BS DL transmission
T_{prop_BS2BS} = propagation delay from between BS-A and BS-B

Figure 4.17 BS2BS-DL2UL interference.

(2) BS2BS interference in UL–DL switching (**BS2BS-UL2DL**), see Figure 4.18. Consider that BS-A timing is ahead of (i.e. earlier than) BS-B timing, then BS-A (aggressor) early DL transmission can interfere with BS-B (victim) end of UL reception.
To avoid BS-A start of DL transmission (point X) overlapping with BS-B end of UL transmission (point Y), we have the constraint in Eq. (4.13):

$$TA_{offset} > T_{sync} + T_{BS\ off\text{-}>\ on} - T_{prop\_BS2BS}$$

(4.13)

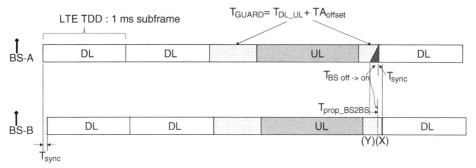

$$T_{GUARD} = T_{DL\_UL} + TA_{offset}$$

T_{GUARD} = TDD switching gap
T_{sync} = difference in BS-A and BS-B synchronization alignment
$T_{BS\ off\text{-}> on}$= transition time to start turning on BS DL transmission
T_{prop_BS2BS}= propagation delay from between BS-A and BS-B

Figure 4.18 BS2BS-UL2DL interference.

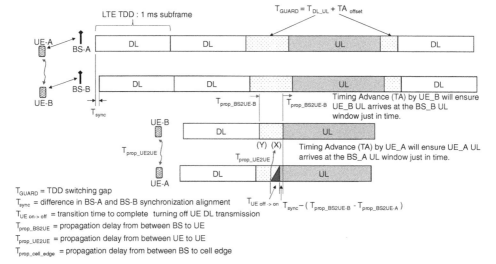

T_{GUARD} = TDD switching gap
T_{sync} = difference in BS-A and BS-B synchronization alignment
$T_{UE\ on\text{-}> off}$ = transition time to complete turning off UE DL transmission
T_{prop_BS2UE} = propagation delay from between BS to UE
T_{prop_UE2UE} = propagation delay from between UE to UE
$T_{prop\_cell\_edge}$ = propagation delay from between BS to cell edge

Figure 4.19 UE2UE-DL2UL interference.

(3) UE2UE interference in DL–UL switching (**UE2UE-DL2UL**), see Figure 4.19. UE-A follows BS-A timing and UE-B follows BS-B timing. Consider that BS-A timing is ahead of (i.e. earlier than) BS-B timing, then UE-A (aggressor) early UL transmission can interfere with UE-B (victim) end of DL reception.

To avoid UE-A start of UL transmission (point X) overlapping with UE-B end of DL reception (point Y), we have the constraint in Eq. (4.14):

$$
\begin{aligned}
&\left(T_{GUARD} - TA_{offset}\right) > T_{sync} - \left(T_{prop\_BS2UE\text{-}B} - T_{prop\_BS2UE\text{-}A}\right) + T_{UE\ off \to on} \\
&+ 2{*}T_{prop\_B2SUE\text{-}B} - T_{prop\_UE2UE} \left(T_{GUARD} - TA_{offset}\right) > T_{sync} \\
&+\left(T_{prop\_BS2UE\text{-}A} + T_{prop\_BS2UE\text{-}B}\right) + T_{UE\ off \to on} - T_{prop\_UE2UE}
\end{aligned}
\tag{4.14}
$$

The worst-case criteria happens when

- Both cells have the same cell size = $T_{prop\_cell\_edge}$
- UE at the cell edge: $T_{prop\_BS2UE-A} = T_{prop\_BS2UE-B} = \alpha_{NLOS} *T_{prop\_cell\_edge}$, where $\alpha_{NLOS} > 1$ as the non-line-of-sight factor)
- UE-A and UE-B are co-located: $T_{prop\_UE2UE} = 0$

The constraint is presented in Eq. (4.15):

$$(T_{GUARD} - TA_{offset}) > T_{sync} + \alpha_{NLOS}*2*T_{prop\_cell\_edge} + T_{UE\ off \to on} \qquad (4.15)$$

(4) UE2UE interference in UL–DL switching (**UE2UE-UL2DL**), see Figure 4.20. UE-A follows BS-A timing and UE-B follows BS-B timing. Consider BS-B timing is behind (i.e. later than) BS-A timing, then UE-B (aggressor) late UL transmission can interfere with UE-A (victim) start of DL reception.

To avoid UE-B end of UL transmission (point X) overlapping with UE-A start of DL reception (point Y), we have the constraint in Eq. (4.16):

$$TA_{offset} > T_{sync} - (T_{prop\_BS2UE-B} - T_{prop\_BS2UE-A}) + T_{UE\ on \to off} + T_{prop\_UE2UE} \qquad (4.16)$$

The worst-case criteria happens when

- Both cells have the same cell size = $T_{prop\_cell\_edge}$
- Both UEs at the cell edge: $T_{prop\_BS2UE-A} = T_{prop\_BS2UE-B} = T_{prop\_cell\_edge}$

The constraint is presented in Eq. (4.17):

$$TA_{offset} > T_{sync} + T_{UE\ on \to off} + T_{prop\_UE2UE} \qquad (4.17)$$

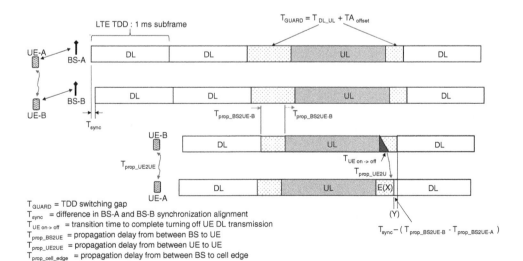

Figure 4.20 = TDD switching gap
T_{sync} = difference in BS-A and BS-B synchronization alignment
$T_{UE\ on\text{-}> off}$ = transition time to complete turning off UE DL transmission
T_{prop_BS2UE} = propagation delay from between BS to UE
T_{prop_UE2UE} = propagation delay from between UE to UE
$T_{prop\_cell\_edge}$ = propagation delay from between BS to cell edge

Figure 4.20 UE2UE-UL2DL interference.

Based on the analysis of each TDD interference type, Table 4.6 summarizes the constraint equations.

To help the reader observe the impact of all the contribution factors in the above constraint equations, 3GPP contribution R4-1703013 [21] suggests some simplifications to combine the relationship.

4.6.2.1 First Simplification

In a first simplification, we only consider co-located UEs as the worst-case condition so that $T_{prop\_UE2UE} = 0$.

As a result of this simplification, to satisfy both Eqs (4.12) and (4.13), we can substitute Eq. (4.13) with $T_{prop\_BS2BS} = 0$ into Eq. (4.12) to form Eq. (4.18):

$$T_{GUARD} > 2{*}T_{sync} + \left[T_{prop\_BS2BS} + T_{BS\ on} \to off \right] + \left[T_{BS\ off} \to on \right] \qquad (4.18)$$

Similarly, to satisfy both Eqs (4.15) and (4.17), we can substitute Eq. (4.17) into Eq. (4.15) to form Eq. (4.19):

$$T_{GUARD} > 2{*}T_{sync} + \left[\alpha_{NLOS} {*} 2{*} T_{prop\_cell\_edge} + T_{UE\ off \to on} \right] + \left[T_{UE\ on \to off} \right] \qquad (4.19)$$

Finally, to satisfy both Eqs (4.18) and (4.19), we can regroup to form Eq. (4.20):

$$T_{GUARD} \geq 2{*}T_{Sync} + \max \left[(T_{BS\ on \to off} + T_{prop\_BS2BS}), \right.$$
$$\left. (T_{UE\ off \to on} + \alpha_{NLOS} {*} 2{*} T_{prop\_cell\_edge}) \right] + \max \left[(T_{BS\ off \to on}), (T_{UE\ on \to off}) \right] \qquad (4.20)$$

Table 4.6 Summary of constraint equations for different interference types.

Interference type	Constraint equation	Worst-case condition
eNB–eNB interference (DL–UL switching) **BS2BS-DL2UL**	Equation (4.12) $(T_{GUARD} - TA_{offset}) > T_{sync} + T_{BS\ on \to off} + T_{prop\_BS2BS}$	$T_{prop\_BS2BS} = ISD = 3{*}Rs$
eNB–eNB interference (UL–DL switching) **BS2BS-UL2DL**	Equation (4.13) $TA_{offset} > T_{sync} + T_{BS\ off \to on} - T_{prop\_BS2BS}$	Worst case is co-located BS so that $T_{prop\_BS2BS} = ISD = 0$
UE–UE interference (DL–UL switching) **UE2UE-DL2UL**	Equation (4.15) $(T_{GUARD} - TA_{offset}) > T_{sync} + \alpha_{NLOS} {*} 2 {*} T_{prop\_cell\_edge} + T_{UE\ off \to on}$	Worst case is when • co-located UE ($T_{prop\_UE2UE} = 0$) • same cell radius (both cells have the same $T_{prop\_cell\_edge} = 2{*}Rs$)
UE–UE interference (UL–DL switching) **UE2UE-UL2DL**	Equation (4.17) $TA_{offset} > T_{sync} + T_{UE\ on \to off} + T_{prop\_UE2UE}$	Although the worst case is when both UEs are close to their own BS with larger ISD, this interference is weak due to limited UE transmission power. The worst case is actually co-located BS with UEs close to their own BS, this means $T_{prop\_UE2UE} = 0$

4.6.2.2 Second Simplification

From Eq. (4.20), we only consider $\alpha_{NLOS} = 1$ (which means BS to UE as line-of-sight propagation in the model). This simplification is optimistic but should be acceptable for this basic analysis.

In addition, we further assume $T_{prop\_BS2BS} = T_{ISD} = 1.5*T_{prop\_cell\_edge}$ (which is $<2*T_{prop\_cell\_edge}$) in Eq. (4.20). This simplification means there is no significant BS–BS interference after passing the first neighbouring cell. This is generally acceptable even if it is not necessarily true for some deployments.

As a result of this simplification, we arrive at the final Eq. (4.21) that can highlight the trade-off clearly. When the final equation is satisfied, it is also guaranteed to satisfy Eqs (4.12) to (4.17):

$$T_{GUARD} \geq 2*T_{Sync} + 2*T_{prop\_cell\ edge} + \max\left[(T_{BS\ on\rightarrow off}), (T_{UE\ off\rightarrow on})\right]$$
$$+ \max\left[(T_{BS\ off\rightarrow on}), (T_{UE\ on\rightarrow off})\right] \tag{4.21}$$

4.6.2.3 Interpretation: The Final Equality

The switching gap (T_{GUARD}) can be configured to be 1 or more symbol duration. To minimize system overheads, the switching gap should be kept as small as possible (1 symbol gap to be ideal). To avoid any type of TDD interference, the achieved ISD budget (which is $1.5*T_{prop\_cell\_edge}$) will be bounded by the switching gap and the chosen synchronization time alignment error budget and BS/UE switching budget. When deployment requires larger cell coverage (i.e. larger cell radius and ISD), one trade-off will be an increase in the switching gap time which leads to more system overheads.

3GPP has evaluated each factor budget carefully based on cost and technology trade-offs. Based on the currently agreed 3GPP NR specification in Release 15 (TS 38.104), the following specifications are stated to drive the 5G synchronization TAE requirement as well as the BS/UE switching time requirement:

- $T_{sync} = 3\ \mu s$, which leads to $\pm1.5\ \mu s$ TAE for each BS site
- $T_{BS\_switch} = 10\ \mu s$ (on$\rightarrow$off or off$\rightarrow$on) for FR1 range, 5 μs for FR2 range
- $T_{UE\_switch} = 10\ \mu s$ (on$\rightarrow$off or off$\rightarrow$on) for FR1 range, 5 μs for FR2 range

It is worth noticing that the agreed T_{sync} specification is not tighter than the LTE-TDD specification. This strategy allows reuse of many existing synchronization solutions, which can be very important for LTE to 5G migration scenarios. The switching time specification is tighter than the LTE-TDD specification but not severely overtightened based on achievable and reasonable cost technology. Using these specifications, the achieved cell radius (or ISD between sites) can be derived based on the selected 5G SCS configuration.

Table 4.7 shows a few examples of SCS configurations and their achieved cell radius based on these 3GPP-agreed contribution factor values. A key observation is the achievable cell radius in the FR1 range (a few kilometres) vs FR2 range (200 m). For the FR1 range, it is a reasonable result since it is more important for the FR1 range to provide blanket coverage similar to the LTE system footprint, especially for migration reasons. For the FR2 range, it is a newer spectrum not used by the LTE system, therefore there is less migration consideration. Instead, a smaller cell radius as trade-off allowing higher carrier bandwidth

Table 4.7 Maximum theoretical distance cell edge for different SCS and 3GPP-agreed contribution factor values.

Band as defined by 3GPP NR spec	FR1 range		FR2 range	
SCS configuration	15 kHz	30 kHz	60 kHz	120 kHz
T_{sync} (µs)	3.0	3.0	3.0	3.0
Symbol duration (µs)	71.4	35.7	17.9	8.9
T_{GUARD} (#symbols)	1	1	1	2
T_{BS_switch} or T_{UE_switch} (µs)	10.0	10.0	5.0	5.0
$T_{prop\_cell\_edge}$ (µs)	22.7	4.9	0.9	0.9
Maximum theoretical distance cell edge (m) ($T_{prop\_cell\_edge}$)	6813.0	1456.5	278.3	278.3

Note: In the 3GPP NR definition, FR1 range (lower frequency spectrum <6 GHz) and FR2 range (centimetre wave and millimetre wave spectrum).

and lower latency using higher SCS configuration is generally acceptable. As a result, a 200 m cell radius range is also reasonable, especially for the FR2 range target application as urban and/or hotspot site.

In conclusion, the current 3GPP NR standard requires the 5G synchronization maximum time alignment error to be 3 µs between any two sites. This translates to max ± 1.5 µs per site TAE requirement as standard practice. Given that TDD technology will be widely deployed in the 5G system in many world markets, this requirement should be taken as the mandatory requirement for any 5G system synchronization target. Later, different 5G synchronization solutions will be introduced to meet this mandatory requirement.

4.6.3 Time Alignment Error (Due to MIMO)

Based on the 3GPP NR definition (TS 38.104), the MIMO accuracy is:

TAE = max 65 ns between any two paths forming the MIMO transmission

This specification is applicable at FR1 range (lower frequency spectrum <6 GHz) and FR2 range (centimetre wave and millimetre wave spectrum). In addition, this specification is defined at a specific interface reference point based on radio/antenna (or antenna array) configuration. Under TS 38.104, we have the following.

- BS Type 1-C: FR1 range, spec reference point is BS antenna connector
- BS Type 1-H: FR1 range, spec reference point is transceiver boundary connector (TAB) of antenna array
- BS Type 1-O: FR1 range, spec reference point is OTA interface of antenna array
- BS Type 2-O: FR2 range, spec reference point is OTA of antenna array

This specification of 65 ns is for any BS type (1-C, 1-H, 1-O or 2-O).

4.6.4 Time Alignment Error (Due to Carrier Aggregation)

Based on the 3GPP NR definition (TS 38.104), there are currently different sets of specifications based on range (FR1 or FR2) and its corresponding radio/antenna (or antenna array) configuration.

Refer to the BS type (1-C, 1-H, 1-O, 2-O) specification reference point definition as shown above.

The specification in Table 4.8 is applicable at the FR1 range interface (1-C and 1-H) for any pair of signals forming aggregated carriers.

The specification in Table 4.9 is applicable at the FR1 range interface (1-O) and FR2 range interface (2-O) for any pair of signals forming aggregated carriers.

Note that the time alignment error for the TDD interference constraint is 3 μs between any sites, which translates to an absolute time alignment error of ±1.5 μs for each site relative to a global reference like GNSS (or GPS). In comparison, the carrier aggregation specification of 130 ns (or 260 ns) is tighter but it is a relative time alignment error. It only applies to the group of signals responsible for the aggregated carriers. Therefore, when this group of signals are formed by a cluster of transmission assets (e.g. multiple RUs), the time alignment error is only among these assets to a common reference. The common reference does not need to be a global reference like GNSS (or GPS) to provide absolute alignment. Between one RU cluster and another RU cluster, there is no alignment requirement. The cluster concept is important in realization since it is generally easier to achieve high relative accuracy among assets within a cluster than to achieve high absolute accuracy across all assets.

Table 4.8 5G system TAE for BTS type 1-C and 1-H.

FR1 range (1-C, 1-H) SCS for data [kHz]	Intraband (contiguous) TAE (ns)	Intraband (non-contiguous) TAE (μs)	Interband TAE (μs)
15	[260]	[3]	[3]
30	[260]	[3]	[3]
60	[260]	[3]	[3]

Table 4.9 5G system TAE for BTS type 1-O and 2-O.

FR1 range (1-O) SCS for data [kHz]	Intraband (contiguous) TAE (ns)	Intraband (non-contiguous) TAE (μs)	Interband TAE (μs)
15	[260]	[3]	[3]
30	[260]	[3]	[3]
60	[260]	[3]	[3]
FR2 range (2-O) SCS for data [kHz]	**Intraband (contiguous) TAE (ns)**	**Intraband (non-contiguous) TAE (μs)**	**Interband TAE (μs)**
60	[130]	[260]	[3]
120	[130]	[260]	[3]

4.6.5 Time Alignment Accuracy (Due to Other Advanced Features)

From LTE experience, there are many other advanced features supported by the 3GPP standard without specific time alignment accuracy specified under the 3GPP standard. The intention of 3GPP is to leave the specification to the solution provider as a proprietary specification to meet the solution provider's feature performance.

For example, these advanced features include:

- CoMP (Coordinated MultiPoint transmission) as a technique to improve cell-edge data rate
- OTDOA (Observed Time Delay of Arrival) as a technique to meet geolocation (North America E911 capability) mandate

Moving into the 3GPP NR standard, the current expectation on 3GPP is to follow the same guidelines. Regardless, these features need time alignment among different sites. Since the general view is no intervendor operability on these features, each vendor solution can define its own time alignment error target to meet its own feature performance. Based on the general nature of these features, time alignment is critical and challenging to ensure good performance. For OTDOA, a much better than ± 1.5 µs absolute TAE per site is expected. For CoMP, a cluster concept may still be possible and a relative 200 ns TAE specification within a cluster (on a par with the carrier aggregation specification) is expected.

References

1 Holma, H., Toskala, A., and Nakamura, T. (2020). *5G Technology: 3GPP New Radio*. John Wiley & Sons.

2 Metsälä, E. and Salmelin, J. (2016). *LTE Backhaul*. John Wiley & Sons.

3 O-RAN.WG4.CUS.0-v07.00. Control, user and synchronization plane specification.

4 IEEE802.1CM Time-Sensitive Networking for Fronthaul.

5 IEEE 802.1AX-2008 Link Aggregation.

6 IEEE802.3cd Media Access Control Parameters for 50 Gb/s and Physical Layers and Management Parameters for 50 Gb/s, 100 Gb/s, and 200 Gb/s Operation.

7 IEEE802.3ck Media Access Control Parameters for 50 Gb/s and Physical Layers and Management Parameters for 50 Gb/s, 100 Gb/s, and 200 Gb/s Operation.

8 3GPP TS38.321 Medium Access Control (MAC) protocol specification.

9 ITU-R report M.2410 Minimum requirements related to technical performance for IMT-2020 radio interface(s).

10 3GPP TS 38.322 Radio Link Control (RLC) protocol specification.

11 3GPP TS 22.261 Service requirements for the 5G system.

12 3GPP TS 22.104 Service requirements for cyber-physical control applications in vertical domains.

13 Høyland, A. and Rausand, M. (1994). *System Reliability Theory, Models and Statistical Methods*. John Wiley & Sons.

14 ITU-T G.911: Parameters and calculation methodologies for reliability and availability of fibre optic systems.

15 Salmela, O. (2005). *Reliability Assessment of Telecommunications Equipment*. Helsinki University of Technology.

16 NGMN: 5G Extreme requirements: End-to-end considerations.

17 3GPP TS 38.300 NR; NR and NG-RAN Overall description.

18 3GPP TS 23.501 System architecture for the 5G system (5GS).

19 3GPP TS 33.210 Network Domain Security (NDS); IP network layer security.

20 3GPP TS 38.211 Physical channels and modulation.

21 3GPP TS 38.104 Base Station (BS) radio transmission and reception.

22 3GPP R4-1703013 TDD timing budget, technical contribution to TR 38.803.

5

Further 5G Network Topics

Esa Malkamäki, Mika Aalto, Juha Salmelin and Esa Metsälä

5.1 Transport Network Slicing

5.1.1 5G System-Level Operation

Network slicing is one of the 5G-introduced new features, requiring standalone (SA) mode to allow UE selection of a specific slice. Slicing is an end-to-end solution, covering the air interface, User Equipment (UE), 5G radio network, 5G core network and transport.

Slice is indicated by Slice/Service Types (SSTs) defined in 3GPP 23.501 [1] (see Table 5.1). SST is an 8-bit field, including an operator-specific value range of 128 to 255.

Slice Differentiatior (SD), a 24-bit field, separates different slice instances of the same slice type. The combination of slice/service type and slice differentiator is called S-NSSAI (Single-Network Slice Selection Assistance Information; see Figure 5.1).

S-NSSAI is provided in the UE context setup signalling and that UE context (bearer) then becomes slice specific, and vice versa, traffic flows of different slices use different bearers. A single UE can have traffic flows on different bearers belonging to different slices.

The 5G RAN part of the system is aware of the slice and can use slice information for differentiated treatment in radio scheduling, air interface L1/L2 configurations, admission control and other functions. These topics are implementation specific within the 5G RAN.

5.1.2 Transport Layers

S-NSSAI itself is not directly visible for the transport layers, so slice indication needs to be given by mapping the S-NSSAI received in radio network signalling to some field that networking devices can read and use for routing, Quality of Service (QoS), etc. This is, of course, necessary only if transport is to provide differentiated treatment for the slice-specific traffic. If that is not the case, then the transport network layer remains agnostic to slicing.

An F1-AP signalling example is given in Figure 5.2, showing S-NSSAI being carried in the UE context setup message between the gNB-CU and gNB-DU. S-NSSAI is in a similar way carried also in NG-AP, so it is available for the RAN nodes in the Protocol Data Unit (PDU) session setup phase.

5G Backhaul and Fronthaul, First Edition. Edited by Esa Markus Metsälä and Juha T. T. Salmelin.
© 2023 John Wiley & Sons Ltd. Published 2023 by John Wiley & Sons Ltd.

Table 5.1   SST.

Slice/service type	Description	Value
eMBB	Enhanced mobile broadband	1
URLLC	Ultra-reliable low-latency communications	2
MioT	Massive Internet of Things (IoT)	3
V2X	Vehicle to everything	4
HMTC	High-performance machine-type communications	5

S-NSSAI

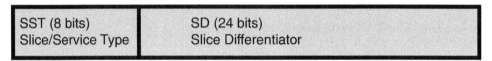

Figure 5.1   S-NSSAI.

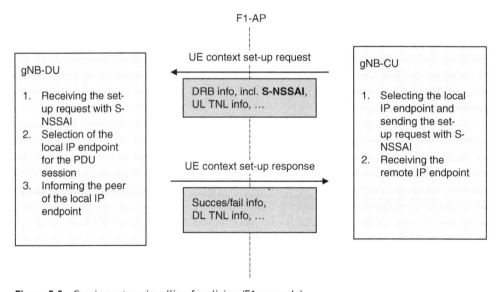

Figure 5.2   Session setup signalling for slicing (F1 example).

In the simplified F1 signalling example of Figure 5.2, the gNB-CU requests the UE context to be set up for a specific slice from the gNB-DU. The setup request messages include, among other things, the list of DRBs (Data Radio Bearers) to be set up, S-NSSAI and also information on the local IP endpoint the gNB-DU should use in the uplink direction. F1-AP signalling is defined in TS 38.473 [2].

The gNB-DU responds by informing success or failure in setup of the DRBs and also tells the local IP endpoint which the gNB-CU should use for the downlink direction data flow.

By means of the slice-specific IP endpoint, the PDU session that is carried over IP packets in the transport network can be logically separated from IP packets belonging to other slices (or to no specific slice), as networking devices (provided with an appropriate configuration) associate a specific IP address or subnet to a specific slice. The same type of configuration is also needed in the mobile network element. The slice-specific IP endpoints then serve, for example, as attachment circuits to connectivity services. These services can be IP Virtual Private Networks (VPNs), as in the example of Figure 5.3, but they could also be something else. The way the slice information is indicated at the transport layer is, in the end, implementation specific. Other methods can be used too.

The setup can fail, as an example, if there are no resources available for the slice-specific PDU session as requested. So there may be further steps required, like admission control and resource checks, before a positive response message can be sent back. Failing in the allocation of resources can be indicated as a cause of failure in the response message from the Distributed Unit (DU) to the Central Unit (CU).

The UE in the upper left corner of Figure 5.3 has two PDU sessions, one in slice A and the other in slice B. The other UE has a single PDU session belonging to slice C. Different VPNs are provided for slices A, B and C using IP MPLS VPN services, where different gNB IP endpoints serve as attachment circuits. Virtual Local Area Networks (VLANs) can be used too. Connectivity is now restricted so that customers (PDU sessions of different slices) in different IP VPNs do not reach each other, so slices are logically isolated.

Additionally, the gNB needs connectivity for common functions like the gNB control plane and network management, and also for non-slice-specific user data flows. These are not shown in Figure 5.3.

Use of an IP VPN is just one possible implementation for logically separated networks with separate connectivity. The alternatives are many: Ethernet services like E-Line, E-Tree or E-LAN can be useful. For harder separation of transport resources, a separate physical link can be allocated for the slice or, for example, a dedicated wavelength in the case of fibre media. The approach of either having shared or dedicated physical links is included in the MEF 84 definitions [3].

If a slice requires bandwidth guarantees, appropriate QoS configurations and GBR (Guaranteed Bit Rate) reservations need to be in place. With IP MPLS VPNs, a path with bandwidth reservations is supported by MPLS TE (Traffic Engineering) services, where

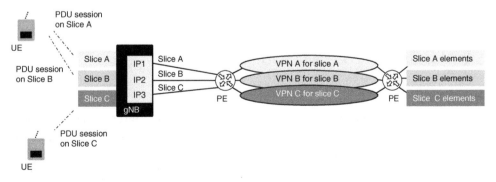

Figure 5.3 Transport network slicing.

resources are reserved from each of the nodes along the path. Segment routing can also provide for limited traffic engineering services.

In case of dedicated physical links, an additional physical link would be needed for the gNB, which in many cases is not feasible economically, especially for sites and markets where fibre availability is limited. Wireless transport can be used as the other link in addition to fibre, as a microwave radio-type solution or IAB (Integrated Access and Backhaul). Nevertheless, dedicated transport links for slices means losing the cost benefits of common shared transport network resources.

QoS is just one possible differentiating factor on the transport network slicing. Security is a must on all logical interfaces, yet there could be differences in cryptographic algorithms, types of services provided, the scope of protection on different transport network layers or use of specified, dedicated nodes.

While there can be many slices at the 5G system level, not all of these necessarily need separated transport resources, not as separate transport services or as separate physical links, and all these slices can be mapped to a common transport service instance. In addition to this single transport service for multiple slices, separate transport services may be provisioned for those slices that require either local processing (local User Plane Function – UPF and server), guaranteed bandwidth or other specific characteristics. This simplifies the design compared to the case where each network slice would be served by a dedicated transport network service, as there are fewer separate transport services to be managed.

The operational burden increases with amount of separate transport services needed. As in Figure 5.3, for the logical separation, when a slice in the gNB access transport is separated by means of IP addresses and/or subnets/VLANs, each slice adds to the number of IP addresses needed for the gNB, which may be an issue at least with IPv4; but even if many IP addresses can be allocated, it still leads to additional operational effort. Furthermore, the transport connectivity service for the slice needs to be provisioned on the transport network side too, be this service an IP VPN, Ethernet service or optical layer path. On top of this, specific QoS features may need to be configured. This may all in practice put an upper limit on the number of slices that can reasonably be supported in the transport network, although, for example, IP VPNs scale very well both in terms of adding new customers and adding new sites for existing customers.

Since slicing is built on the end-to-end concept, administrative boundaries may easily be crossed on the transport path from the access to the core. Efficient management tools that work over multiple domains are essential in managing this type of network.

5.2 Integrated Access and Backhaul

5.2.1 Introduction

Building of new 5G networks, especially on new high-frequency mmWave bands, requires much denser cell sites than in lower-frequency bands typical of 4G networks, since the attenuation of the radio signal is much higher at high frequencies. Furthermore, mmWave deployments typically require almost line-of-sight (LOS) propagation since the high frequencies do not penetrate through buildings and other obstacles. Therefore, the initial 5G mmWave deployments will require new fibre or wireless backhaul solutions to meet the required smaller intersite distances. Here, wireless backhauling via IAB can provide a faster and more cost-efficient way for (initial) 5G deployments.

Wireless backhauling with IAB-nodes can be used mainly for coverage enhancement to mitigate sparse fibre availability, remediate isolated coverage gaps (e.g. behind buildings or trees) and bridge coverage from outdoor to indoor (e.g. wall or window-mounted IAB-nodes for fixed wireless deployments).

Due to the LOS propagation requirement for mmWave deployments, a typical urban grid environment would require fibre connection to almost every street corner and, due to higher attenuation of the mmWaves, even single-street coverage typically requires a base station in almost every lamp post. Typically, operators do not have fibre available everywhere, and wireless backhauling with IAB can significantly improve the coverage of mmWave deployments.

Increasing the density of base stations can also enhance the capacity of the radio access network, especially by providing better radio quality and, therefore, increased data rates. Thus, IAB-nodes can also enhance the capacity of the radio access network but when in-band relaying is used for backhauling, the capacity enhancements are typically limited since the same data needs to be transmitted in addition to the access link over the backhaul link or links in case of multi-hop.

In 3GPP Release 10, relay operation was specified for LTE-Advanced but LTE relays were not deployed widely. LTE relays were L3 relays where the relay node implemented the full 3GPP radio protocol stacks and only a single relay hop was supported. IAB uses L2 relaying where only the lower layers (PHY, MAC, RLC) of the radio protocol stack are used for backhauling. Furthermore, multi-hop relaying is supported by IAB.

IAB is specified in 3GPP Release 16; Stage 2 is described in 3GPP TS 38.300 and 3GPP TS38.401 [4, 5].

5.2.2 IAB Architecture

IAB supports wireless relaying for 5G. The relay node is called the IAB-node and it supports both access and backhaul using NR radio access (see Figure 5.4). The network node terminating the wireless backhauling on the network side is called the IAB-donor gNB. The IAB-donor gNB is a gNB with added functionality to support IAB. IAB wireless backhauling supports both single and multiple hops.

The IAB architecture takes advantage of the split gNB architecture with the CU in the IAB-donor and the DU in the IAB-node. For the IAB-node, the F1 interface specified between gNB-CU and gNB-DU is extended over the wireless backhaul connecting the DU in the IAB-node (IAB-DU) and the CU in the IAB-donor (IAB-donor-CU). The core network interface (NG interface) terminates at the IAB-donor-CU. Thus, the IAB-node is a radio access network node with limited visibility to the core network.

The IAB-DU acts like any gNB-DU terminating the NR access interface to UEs. In addition, the IAB-DU terminates the backhaul link of the next-hop IAB-nodes.

The IAB-node also supports UE functionality, which is referred to as IAB-MT (Mobile Termination) in RAN specs (38.xxx) or IAB-UE in SA specs 3GPP TS 23.501 [1]. IAB-MT supports, for example, a physical layer, layer-2, RRC (Radio Resource Control) and NAS (Non-Access Stratum) functionality, and IAB-MT connects the IAB-node to IAB-DU of another IAB-node (multi-hop) or to an IAB-donor. Furthermore, IAB-MT connects to an RRC layer in IAB-donor-CU and a NAS layer in the Access and Mobility Management Function (AMF).

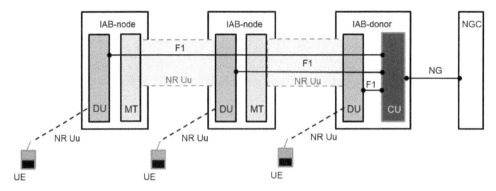

Figure 5.4 IAB architecture and functional split.

5.2.3 Deployment Scenarios and Use Cases

IAB-nodes that are connected to an IAB-donor via single or multiple hops form either a spanning tree topology (IAB Mobile Termination (IAB-MT) connected to a single parent node) or a Directed Acyclic Graph (DAG) topology (IAB-MTs can be dual connected to two parent nodes), see Figure 5.5. In 3GPP Release 16 specs, the IAB-MT can be either single or dual connected (i.e. more than two parents for a given IAB-node are not supported).

IAB supports relaying for both SA (i.e. NR only) and non-SA (NSA, i.e. LTE and NR dual connectivity) deployments, see Figure 5.6. In the SA case, both access UEs and IAB-MTs use NR only and both connect to the 5G core. For NSA, there are two variants, which are both supported in Release 16:

- Access UEs are NSA UEs and connected to the EPC (Evolved Packet Core), whereas IAB-MTs are NR only (SA) and connected to the NGC (Next Generation Core).
- Both access UEs and IAB-MTs support NSA and both are connected to the EPC.

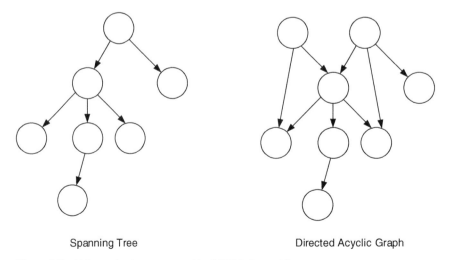

Spanning Tree   Directed Acyclic Graph

Figure 5.5 IAB topologies supported in 3GPP Release 16.

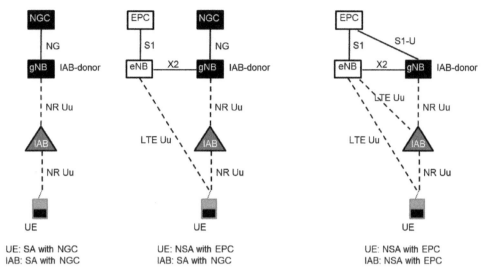

Figure 5.6 IAB deployments: SA (NR only) and NSA (LTE/NR dual connectivity).

The Release 16 specification supports stationary IAB-nodes only (i.e. fixed site installations are assumed and IAB-nodes are not moving). Due to mmWave blockage, the backhaul link may sometimes deteriorate and, therefore, Release 16 IAB supports topology adaptation where the backhaul connection of an IAB-node can be switched from a parent to another parent. Support for mobile IAB-nodes is planned for Release 18 and beyond.

Both in-band (access and backhaul on the same frequency) and out-of-band (backhaul on a different frequency) backhauling is supported. From a specification point of view, IAB can be used on both FR1 (<6 GHz) and FR2 (mmWave), but in practice IAB is best suited for FR2 as discussed in the previous section.

IAB supports NR relaying for NR access only in Release 16. NR relaying for LTE access will be studied in later releases. There are some security issues that need to be solved.

5.2.4 IAB Protocol Stacks

The user plane protocol stack for IAB is shown in Figure 5.7. It shows the L2 relaying, where backhauling uses NR PHY, MAC and RLC layers as well as a new adaptation layer called Backhaul Adaptation Protocol (BAP). F1-U between the IAB-DU and the IAB-donor-CU is carried over IP (i.e. F1-U to IAB-DU is, from a protocol stack point of view, seen as a standard F1-U interface to any gNB-DU specified in 3GPP TS38.470). The full F1-U stack (GTP-U/UDP/IP) is carried on top of the BAP layer on the wireless backhaul. BAP enables backhaul routing over multiple hops in the IAB tree. IP is terminated in the access IAB-node in a similar way as IP is terminated in a normal gNB-DU.

BAP PDUs are carried by backhaul (BH) RLC channels. Multiple BH RLC channels can be configured for each BH link to allow traffic prioritization and QoS enforcement. A BH RLC channel is configured separately for each hop (i.e. the BH RLC channel is a single-hop channel and thus hop-by-hop ARQ is supported by RLC). The UE bearers terminate in the

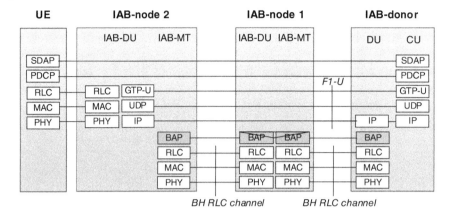

Figure 5.7 User plane protocol stacks for UE, IAB-nodes and IAB-donor.

IAB-donor-CU (i.e. SDAP – Service Data Adaptation Protocol and PDCP – Packet Data Convergence Protocol are between the access UE and the IAB-donor-CU).

The IAB-donor-DU maps the received GTP-U/UDP/IP packets to BH RLC channels and different BH routes based on the destination IP address, DSCP and IPv6 Flow Label.

Control-plane protocol stacks for IAB are shown in Figure 5.8. Like the user plane, a full F1-C stack (F1AP/SCTP/IP) is carried on the BAP layer over the wireless backhaul links. Control-plane traffic can be carried over a dedicated BH RLC channel or multiplexed together with the user plane traffic.

Both F1-U and F1-C are protected via Network Domain Security (NDS) (e.g. via IPsec), as specified in TS 33.501 [6]. The security layer is not shown in the protocols of Figures 5.7 and 5.8. The IP layer on top of the BAP also carries some non-F1 traffic, like signalling for management of IPsec and SCTP (Stream Control Transmission Protocol) associations.

The IAB-MT connects first as a UE and therefore it establishes its own Signalling Radio Bearers (SRBs) for RRC and NAS signalling, as well as Data Radio Bearers (DRBs) (e.g. for

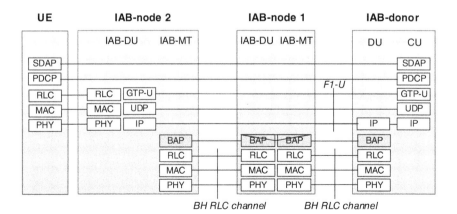

Figure 5.8 Control-plane protocol stacks for UE, IAB-nodes and IAB-donor.

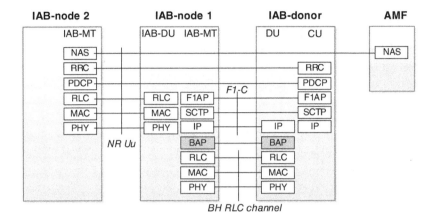

Figure 5.9 Control-plane protocol stacks for the support of IAB-MT SRBs (for RRC and NAS).

Operations, Administration and Management – OAM traffic). The control plane stack of IAB-MT is shown in Figure 5.9. These SRBs and DRBs are transported to the parent-node DU in the same way as normal UE traffic (i.e. there is no BAP layer over the first wireless hop). The parent node encapsulates these bearers into F1-U and F1-C to be carried over the following backhaul hops.

5.2.5 IAB User Plane

5.2.5.1 Backhaul Transport

As discussed in the previous section, both the user plane and the signalling traffic between the IAB-DU and the IAB-donor-CU is carried over the F1 interface. The full F1 stack (GTP-U/UDP/IP for user plane and F1AP/SCTP/IP for control plane) is carried over the wireless backhaul. In addition to F1 traffic, some other IP packets (e.g. SCTP) are carried over the wireless backhaul.

IP traffic to the IAB-DU is routed over the wireless backhaul using the BAP. IP packets that are sent from the IAB-donor-CU to the IAB-node-DU are first IP routed to the IAB-donor-DU, then encapsulated into BAP PDUs in the IBA-donor-DU; BAP PDUs are transmitted over wireless backhaul(s) by using backhaul RLC channels and decapsulated in the IAB-node. Similarly, the IP packets from the IAB-node-DU are encapsulated into BAP PDUs in the IAB-node and the BAP PDUs are decapsulated in the IAB-donor-DU; the IP packets are then IP routed to the IAB-donor-CU.

5.2.5.2 Flow Control

Flow control is used for mitigating congestion in the wireless backhaul. Downstream, the NR user plane protocol, specified in TS 38.425 [7] and used over F1-U, includes end-to-end flow control between the IAB-donor and the access IAB-node serving the access UEs. In addition to that, a hop-by-hop buffer reporting at the BAP layer is specified for DL. Upstream, normal buffer status reporting and UL scheduling grants provide the needed flow control.

5.2.5.3 Uplink Scheduling Latency
In order to reduce the uplink latency, an IAB-node can send a pre-emptive Buffer Status Report (BSR) informing – in addition to the available data amount – also the expected data amount from its child IAB-nodes and access UEs.

5.2.6 IAB Signalling Procedures

In this section we describe some IAB signalling procedures related to IAB topology adaptation, like IAB-node integration, IAB-node migration and IAB topological redundancy procedures, as well as BH RLC channel establishment procedure.

5.2.6.1 IAB-Node Integration
There are four major steps when a new IAB-node connects to an IAB-donor or another IAB-node (multi-hop; see Figure 5.10) [5].

1) An IAB-MT connects to the network as a normal UE and indicates to the network that it is part of an IAB-node. During the IAB-MT connection setup, SRBs are set up for RRC and NAS signalling and a DRB and PDU session for OAM. A new SIB1 parameter is introduced to tell the IAB-MTs which cells support IAB. Also, at least one backhaul RLC channel is set up during this phase.
2) The IAB-donor-CU updates the routing information both downstream and upstream in all related IAB-nodes and IAB-donor-DU via F1AP signalling. The newly added IAB-node is given a BAP address and IP address via RRC signalling, thus creating IP connectivity to the IAB-node.

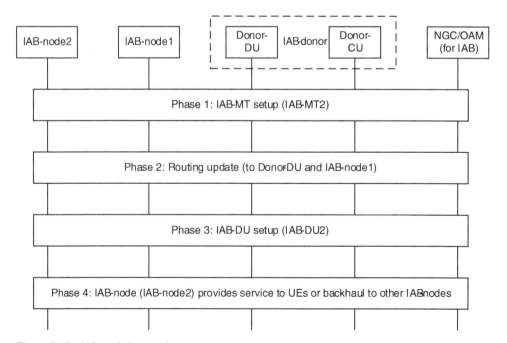

Figure 5.10 IAB-node integration.

3) The IAB-DU performs the F1AP setup procedure. Before F1AP signalling can start, an SCTP and optionally IPsec are set up between the IAB-node and the IAB-donor-CU.
4) The IAB-DU is activated and the IAB-node starts serving UEs and other IAB-nodes.

5.2.6.2 BH RLC Channel Establishment

In order to establish a BH RLC channel between two IAB-nodes or between an IAB-node and IAB-donor-DU, the IAB-donor-CU first sends an F1AP UE Context Setup/Modification Request message to the parent-node DU indicating the BH RLC channel ID and QoS parameters. The parent-node DU sets up the indicated BH RLC channel(s), creates the lower-layer part of the RRC reconfiguration and sends it to the IAB-donor-CU in the F1AP UE Context Setup/Modification Response message. The IAB-donor-CU then sends an RRC Reconfiguration message to the child-node IAB-MT, configuring the BH RLC channel on the IAB-MT side. The above procedure is repeated for each hop when a BH RLC channel needs to be established over multiple hops.

5.2.6.3 IAB-Node Migration (Topology Adaptation) (Intra-CU)

An IAB-node can migrate from a source parent node to a target parent node under the same IAB-donor-CU if, for instance, the backhaul connection to the source parent node deteriorates or due to load balancing. There are three major steps in the IAB-node migration:

1) The migrating IAB-MT makes a handover to the target parent node. The IAB-MT may have provided measurement results to the IAB-donor-CU via RRC signalling. The IAB-donor-CU can then send a handover command as an RRC message to the IAB-MT, which then performs the handover procedure.
2) Routing update: After the IAB-MT has connected to the new parent DU (IAB-DU or IAB-donor-DU), the IAB-donor-CU updates the routing in all related nodes via F1AP signalling. The F1 connection to the migrating IAB-node DU is redirected to this new route. Since the CU does not change, SCTP and IPsec associations do not change and the F1 connection can continue.
3) IAB-DU continues serving UEs. Since the CU does not change, the IAB-DU need not change its cell identifiers (PCI or CGI). Thus, the UEs connected to the IAB-DU are unaware of the topology adaptation.

3GPP Release 16 only supports intra-CU topology adaptation. Inter-CU topology adaptation with partial migration, where only the IAB-MT of the migrating IAB-node connects to the target IAB-donor-CU while the IAB-DU remains connected to the source IAB-donor-CU, will be specified in Release 17. Inter-CU topology adaptation with full migration would require release of the F1 connection and establishment of a new F1 connection to the IAB-DU, since a DU can only be connected to a single CU-CP. This will be specified in Release 18.

5.2.6.4 Topological Redundancy (Intra-CU)

IAB-nodes may have redundant routes to the IAB-donor-CU. An NR dual-connectivity framework is used to enable route redundancy within one IAB-donor-CU, and therefore topological redundancy is only supported in SA mode. The IAB-MT is then configured with

BH RLC channels on two parent radio links. There are three major steps in setting up the topological redundancy:

1) The IAB-MT connected to a parent node (master node) makes measurements and provides them to an IAB-donor-CU via RRC signalling. The IAB-donor-CU may decide to add a secondary leg for an IAB-MT after receiving measurement results from the IAB-MT, and communicates with the secondary parent node to prepare for dual connectivity. The IAB-donor-CU then sends an RRC message to the IAB-MT to set up the secondary connection.
2) BAP routing and BH RLC channel mapping for the related nodes along the added secondary path are configured via F1AP signalling.
3) New TNL address(es) to dual-connecting IAB-node-DU F1-C associations are added if the secondary leg uses different IAB-donor-DU.

Once the dual connectivity is set up, some of the UE bearers connected to the IAB-node can start using the secondary path, and thus the dual connectivity can be used for load balancing.

5.2.7 Backhaul Adaptation Protocol

For IAB, an additional L2 protocol layer called Backhaul Adaptation Protocol (BAP) is introduced in the wireless backhaul links on top of the RLC layer. It supports routing in the IAB topology as well as traffic mapping to BH RLC channels, thus enforcing traffic prioritization and QoS. Details can be found from 3GPP TS 38.340.

5.2.7.1 BAP Routing

For routing in the IAB topology, each BAP PDU header contains a BAP Routing ID which consists of a 10-bit-long BAP address of the destination node and a 10-bit-long Path ID. Based on the BAP address, each node can check whether this BAP PDU should be delivered to the upper layers of this node or forwarded to the next hop node. The BAP address of an IAB-node is configured by the IAB-donor-CU via RRC signalling when a new IAB-node is connected to the IAB topology.

Path ID is used to differentiate multiple routes leading to the same destination node, see Figure 5.11. Different paths can be used for centralized load balancing by mapping UE bearers to different paths. The UE bearer mapping to Path ID is configured by the IAB-donor-CU via F1AP signalling and enforced by the IAB-donor-DU for downlink and by the IAB-node for uplink. Any local load balancing is not supported in Release 16.

The routing configuration is provided to IAB-nodes and the IAB-donor-DU via F1AP signalling. This configuration defines the mapping between the BAP Routing ID (BAP address and Path ID) carried in the BAP PDU header and the BAP address of the next hop node. The routing configuration of a node may contain multiple entries with the same destination BAP address but different Path IDs. In a given node these entries may point to the same or different egress BH links.

Intermediate IAB-nodes can redirect the BAP PDU to another path leading to the same destination BAP address, in case there is an RLF (Radio Link Failure) on the link indicated by the Path ID in the BAP header. Upstream, each IAB-node can be simultaneously connected to two parent nodes using NR dual connectivity between the IAB-MT and the parent-node DUs.

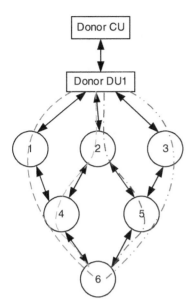

Figure 5.11 Different paths can be configured to a given destination node.

5.2.7.2 BH RLC-Channel Mapping

In the access IAB-node the IAB-donor-CU configures, via F1AP signalling, the mappings between the backhaul RLC channel and the upstream F1 and non-F1 traffic originated in the IAB-node. The mapping is configured separately for each F1-U GTP-U tunnel (TEID), non-UE-associated F1AP messages, UE-associated F1AP messages of each UE and non-F1 packets. Multiple UE bearers can be mapped to a single BH RLC channel (N:1 mapping) or a separate BH RLC channel can be configured for each UE bearer (1:1 mapping). The latter is typically used for GBR bearers.

Similarly for the IAB-donor-DU, the mappings between IP header fields (IP address, DSCP and/or IPv6 Flow Label) and the BH RLC channel ID are configured for downstream traffic by the IAB-donor-CU via F1AP signalling.

In the intermediate IAB-nodes, the traffic mapping from an ingress BH RLC channel to an egress BH RLC channel is configured by the IAB-donor-CU via F1AP signalling. The mapping includes also the BAP addresses of the prior-hop and next-hop nodes.

5.2.8 BH Link Failure Handling

When an RLF happens in an IAB backhaul link, the same mechanisms and procedures are applied as for the access link. The IAB-MT monitors the channel quality of the backhaul link and if an RLF is detected, the IAB-MT initiates recovery using an RRC re-establishment procedure.

If the RRC re-establishment procedure to recover the BH link fails, the IAB-node may transmit an RLF notification message to its child nodes. This notification is sent on the BAP layer using a BAP Control PDU. When a child IAB-node receives an RLF notification message, it considers the BH link where it received the notification as failed (i.e. as if it had detected an RLF on that BH link). When a BH link to a child IAB-node fails, the parent node can use the existing F1AP signalling to inform the IAB-donor-CU about the failure. The IAB-donor-CU can then take action to reroute the F1 connections to the affected IAB-nodes.

In case an IAB-node is configured with NR-DC and only one of its parent links fails, it follows the normal MCG (Master Cell Group) or SCG (Secondary Cell Group) failure procedure to inform the IAB-donor-CU about the failure. RRC re-establishment is avoided in this case and traffic can be rerouted (locally in the node as well as by the IAB-donor-CU) using the still available BH link.

5.2.9 IAB in 3GPP Release 17 and Beyond

In 3GPP, IAB work will continue in the coming releases. For IAB Release 17, enhancements in the following areas will be specified:

- Duplexing enhancements: Release 16 IAB only supports Time-Division Multiplexing (TDM) of radio resources between IAB-MT and IAB-DU. In Release 17, Space-Division Multiplexing (SDM) and Frequency-Division Multiplexing (FDM) support and full duplex capability will be added to PHY specs.
- Resource management enhancements to have full support for dual connectivity in PHY specs.
- Procedures for IAB-node partial migration between IAB-donor-CUs where only the IAB-MT is migrated to another IAB-donor-CU (Release 16 limits this to single IAB-donor-CU scenarios).
- Enhancements to reduce service interruptions due to IAB-node migration and BH RLF recovery (e.g. conditional handover for topology adaptation).

Mobile IAB-nodes were considered for Release 17 but due to the high workload of 3GPP working groups, were postponed to later releases.

5.3 NTN

5.3.1 NTN in 3GPP

3GPP Release 15 introduced a study item for 5G supporting Non-Terrestrial Networks (NTNs) with an aim to understand the deployment use cases, develop channel models, etc. – see TR 38.811 [8]. The work continued with the two following studies in Release 16: TR 23.737 [9] for architecture aspects and TR 38.821 [10] for 5G solutions. In Release 17, two work items were agreed on: 5G over NTN and NB-IoT (narrowband Internet of Things)/eMTC (enhanced Machine-Type Communications) over NTN.

A key driving factor is the lowered cost of launching satellites, especially for the low earth orbits. For 5G, this helps complement coverage in areas that are difficult to reach (mountains or oceans) or uneconomical for terrestrial network rollout (very sparsely populated areas). A further benefit is that with NTNs, service availability is maintained in case of natural catastrophes and similar emergencies.

Compared to terrestrial 5G, with NTNs there are multiple differences, including:

- Long distance to satellite causes large attenuation and difficulties for protocol operations like H-ARQ.
- Performance is impacted by the delay and other factors.
- For Non-Geostationary Orbit (NGSO) satellites, satellite movement causes time drift and Doppler shift.
- NGSO satellite movement also leads to frequent handovers as a terminal is served by different satellites constantly, from a few seconds to a few minutes in the Low Earth Orbit (LEO) scenario.
- The area covered by the satellite is large and moves with the NGSO satellite, unless it is compensated by directing the beam.
- It is difficult (or not feasible in practice) to contain the coverage area to match national borders, yet some means to control the access is needed, as PLMNs (Public Land Mobile Networks) may not be allowed to offer service outside national boundaries.
- With normal terminals performance is limited, one limitation is the uplink power from the UE.

The benefits with the combination of NTN and 5G technologies are that 5G coverage can be extended and, on the other hand, NTN can rely on common 5G technology and, of course, new use cases can be served by the connectivity that 5G NTN offers. With LEO satellites, 3GPP Release 17 5G terminals are usable with limitations in performance. A special Very Small Aperture Terminal (VSAT) is an alternative.

One use case is for IoT terminals reporting a condition or status via the satellite to the cloud, and for this use case a continuous service is not a must. Solutions depend on the targeted use case.

5.3.2 Different Access Types

NTN items as in [9, 10] cover different approaches, from High-Altitude Platform Systems (HAPS) to Geostationary Orbit satellites (GEOs). See Table 5.2. Unmanned Aerial Systems (UAS), e. g. drones, cover altitudes starting from far less than 8 km.

The delay from UE to satellite is on the order of 3–15 ms for LEO (800 km), 27–43 ms for MEO (8000 km) and 120–140 ms for GEO. A 5G system is expected to take the long latency due to NTN into account for QoS. Services that require low latency may not be supported.

The satellite payload consists either of a simpler relay function (transparent, also called 'bent pipe') or it can have a regenerative payload. These alternatives are shown in Figure 5.12.

Table 5.2 Different platforms [10].

Platform		Altitude
High Altitude Platform Systems	HAPS	8–50 km
Low Earth Orbit Satellites	LEO	300–1500 km
Medium Earth Orbit Satellites	MEO	7000–25,000 km
Geostationary Earth Orbit Satellites	GEO	35,768 km
High Elliptical Orbit Satellites	HEO	400–50,000 km

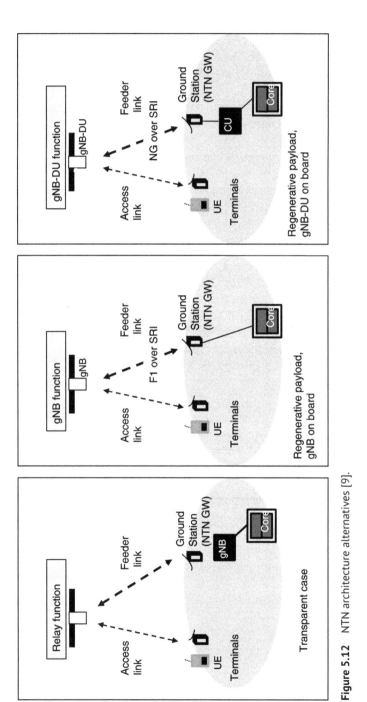

Figure 5.12 NTN architecture alternatives [9].

With the transparent architecture, the whole 5G RAN is located on the ground. The satellite payload includes frequency conversion, relaying the signal from the gNB via the ground station and feeder link to the satellite and then back to the access using a different frequency.

With a regenerative payload there are two alternatives, either gNB-DU or full gNB functionality is located in the satellite payload while the rest of the 5G network is on the ground. The benefit of a regenerative payload is that the distance from the 5G baseband to the ground is shorter, and the latency for L1/L2 protocols is shorter. The satellite payload becomes more complex due to the RAN functionality on board. From the architectures of Figure 5.12, 3GPP has focused in Rel-17 and Rel-18 to the transparent case.

Inter-Satellite Links (ISLs) are included in the regenerative architecture so that a satellite can connect to a ground station via another satellite, or even via multiple ISLs.

In both transparent and regenerative payload architectures, a ground station supports the feeder link to the satellite. The ground station then connects them to further 5G network elements.

For non-geostationary satellites, the movement of the satellites requires connection via different ground stations and also to different mobile network elements on the ground. Terminals also need to connect to the mobile network elements via different satellites.

A further topic is that service is not continuously available if the number of satellites in the constellation is not high enough, a situation that is likely at least in the early phase.

5.3.3 Protocol Stacks

For the transparent case, the NTN elements – NTN gateway (ground station) and satellite – appear as RF converters/frequency shifters and do not terminate 5G protocols (see Figure 5.13).

The whole 5G stack is processed in the gNB at ground, and the role of the NTN GW is to transport the signal to the satellite (with selected frequency); the satellite's role is to relay

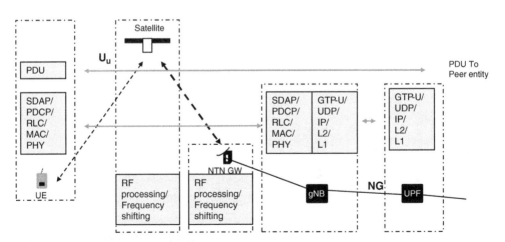

Figure 5.13 Transparent satellite user plane stack [9].

the signal back towards the ground at a 5G air interface frequency. At the terrestrial part, the gNB interfaces the UPF with a standard NG interface. A delay due to the satellite now occurs over the 5G air interface. The control plane stack is similarly carried transparently over the NTN GW and the satellite to the UE.

In the regenerative payload case, gNB-DU or full gNB functionality is placed into the satellite. The gNB-DU case is shown in Figure 5.14 and the gNB case in Figure 5.15. When the satellite hosts the DU, F1 is carried over the Satellite Radio Interface (SRI), and when the satellite hosts the gNB, NG is carried over the SRI. Both interfaces require adjustments and configuration to cope with the latency involved.

At the ground station, NTN GW acts as a router between the interface to the 5G network and the satellite, supporting forwarding of the F1 or NG IP packet over the SRI. The SRI or NTN GW does not terminate 5G protocols but acts as a transport layer device. The SRI protocol layers are not defined by 3GPP.

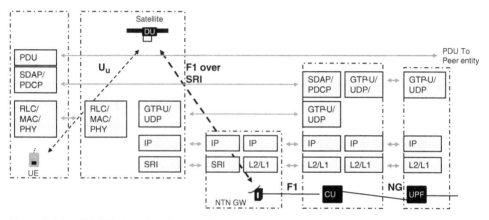

Figure 5.14 gNB-DU placed on the satellite (user plane stack) [9].

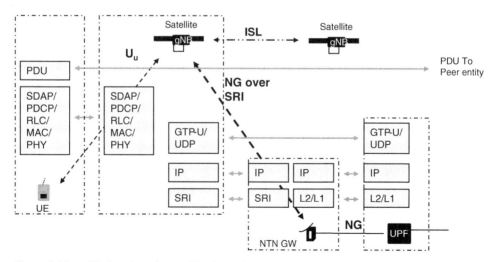

Figure 5.15 gNB placed on the satellite (user plane stack) [9].

NTN GW provides an SRI, on top of which F1 is carried over IP. F1 is terminated in the satellite payload (gNB-DU functionality), and then 5G air interface protocols up to the RLC layer (RLC/MAC/PHY) are hosted by the gNB-DU, as normal. A delay due to the satellite occurs now for the F1 interface and also movement of the satellite concerns F1. The remaining user plane protocols, SDAP and PDCP, are terminated in the gNB-CU, which resides on the ground behind the NTN GW.

The benefit of this approach is that now real-time critical protocols RLC/MAC/PHY are located closer to the UE and there is less delay. The F1 interface has to tolerate a longer delay than usual.

The control plane stack is not shown, but here as well the split follows the functional split between gNB-DU and gNB-CU as in the terrestrial network case.

When gNB functions are located in the satellite, it is the NG interface that is carried over the SRI between the satellite and NTN GW. In this case, full 5G RAN functionality is located in the satellite. The NG interface has to tolerate a longer delay than usual for protocol operation and also satellite movement now impacts NG. Logical interfaces (F1, NG, Xn) additionally need to be protected cryptographically within the network domain as defined by 3GPP (not shown in the protocol stacks).

Comparing the architecture alternatives, the transparent case is the simplest in that there is the least functionality required in the satellite and also all logical interfaces (except the Uu) are located on the ground. The downside is the longer propagation delay between the UE and the baseband, as the signal needs to travel twice the distance from the ground to the satellite. This causes optimization needs for the 5G air interface and for real-time protocol operation, like H-ARQ. With either of the regenerative alternatives, real-time critical protocols are supported within the satellite payload.

It is also possible to support an ISL as shown in Figure 5.15, such that the satellite may connect to the ground via another satellite or multiple satellites and not directly when the ground station is not reachable directly (which might happen e.g. in the middle of an ocean). In that case a path to the ground station may be found through use of one or multiple ISL links.

5.3.4 Transparent Architecture

In 3GPP the transparent architecture has been the first target and several enhancements for this architecture are included in TS38.300 [4]. With a transparent architecture, a gNB logically consists of two parts: an NTN-specific infrastructure system and then the gNB functions as in Figure 5.16.

Logically, the gNB includes not only the gNB functions already defined by 3GPP but also the NTN system, since the air interface in this case is supported from the NTN payload, which is on the airborne/spaceborne vehicle (e.g. a satellite or HAPS). The NTN system includes this airborne/spaceborne vehicle with the payload and then the NTN GW, with a feeder link between the GW and the satellite.

The NTN system and gNB functions both interface an NTN control function and an O&M (Operations and Management) function. The NTN control function is needed for the gNB (in addition to other functions) for the feeder link switchover, which triggers handovers for the NTN access UEs. Otherwise, the NTN control function controls the spaceborn/airborne

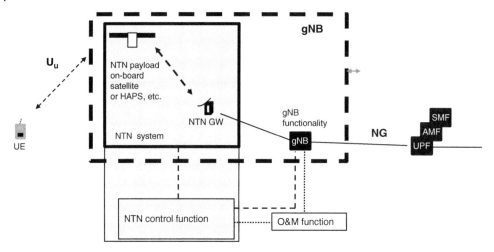

Figure 5.16 gNB in transparent architecture.

vehicle, the feeder link interface, radio resources and the NTN GW. The O&M function provides the gNB with configuration parameters and satellite ephemeris information.

The NG interface towards the core network is a standard one as defined by 3GPP. In the logical gNB of Figure 5.16 there are, however, many NTN-specific interfaces that are out of scope of 3GPP and as such unspecified:

- The interface between the gNB functions and the NTN GW.
- The interface between the NTN control function and the gNB.
- The O&M interface.

Also the feeder link is unspecified in 3GPP. The NTN system is essentially a black box which 'extends' the Uu interface from the local gNB via the airborne/spaceborne vehicle to another area. It is like an RF repeater, with a function where the signal is shifted to a selected feeder link carrier and then the satellite payload shifts the frequency back to the selected 5G band. The operation and management of the NTN payload and GW and radio resources within the NTN system is the functionality of the NTN control function; implementation of that functionality is within the NTN infrastructure.

Indirectly, NTN-specific characteristics become visible within the 5G RAN, due to the topics discussed: long latency, time shift, performance, etc. Also, the core network knows that the PDU session involves an NTN system and can take this information into account in its functions for access and session management.

5.3.5 Feeder Link Switchover

In addition to mobility for the UEs, which is supported in the connected mode by handovers, a new aspect with NTN is that the feeder link also needs to be switched when the satellite moves. In 3GPP, this is specified as a TNL (Transport Network Layer) function and is undefined by 3GPP. Satellite movement is known, so switching can be based on time and location (ephemeris information) or on other methods like reports and instructions

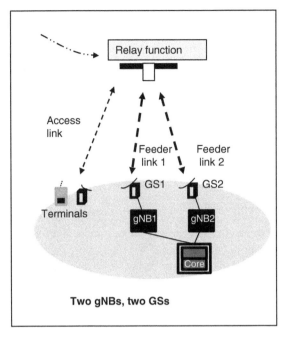

Figure 5.17   Seamless feeder link switchover example [4, 9].

between the satellite, NTN GW and control function. When the feeder link switches and the satellite connects to a new ground station, a handover is caused on the access side so the feeder link switchover and handovers for the UEs occur simultaneously.

The target for the feeder link switchover is that it does not cause a break to the UE communications. The feeder link switchover can be a soft one if the satellite maintains links to two ground stations simultaneously with different radio resources during the switchover (see Figure 5.17).

For the transparent case depicted in Figure 5.17, when moving towards GS2, the satellite maintains feeder link 1 to GS1 and gNB1, while establishing feeder link 2 to gNB2. During the transition to GS2, the satellite reflects radio resources from both gNB1 and gNB2 to the access side and UEs are moved by Xn or NG handover procedures to gNB2 before the connection to GS1 and gNB1 is dropped. The TNL feeder link switchover allows seamless connectivity for the UEs.

5.4   URLLC Services and Transport

5.4.1   Background

In Section 2.10, 3GPP enhancements for URLLC (Ultra Reliable Low Latency Communication) were introduced, starting from Release 15 with further enhancements in Releases 16 and 17. These enhancements are radio and system-centric and as such do not define any specific functionality for the transport links. Transport network optimizations for low latency and high reliability can be found from other standard organizations, like IEEE TSN and IETF DetNet technologies [11, 12].

From the 3GPP angle, a lot of freedom exists on the transport network side. Constraints and pre-requisites are set by URLLC service requirements and the 5G system architecture, and whether specific transport functionality is required depends on these requirements and on the 5G deployment. A specific transport-related area is security; the logical interfaces are cryptographically protected as defined by 3GPP but the deployment needs to be analysed against specific vulnerabilities and threats, since transport links and networks by definition are designed to provide connectivity. With URLLC, many services depend on the network services being safe and trusted.

Dependencies on 5G system functionality exist, as enhancements like PDCP duplication potentially impact and benefit transport links as discussed in Chapter 4, so transport should be considered as part of the system. The flexibility in 5G system architecture means, for the URLLC, that network functions like UPF can be located close to the cell site to shorten any transport link distance and correspondingly cut the amount of propagation delay consumed in transport links, while keeping functions for other (non-URLLC) services on a bigger core site. These examples highlight the dependencies on the 5G system.

Service requirements in 3GPP TS 22.261 [13] point further to other documents: cyber-physical control applications in vertical domains, TS 22.104 [14]; V2X in TS 22.186 [15]; and rail communications in TS 22.289 [16].

Cyber-physical applications are divided into communication needs that are characterized by

- Periodic deterministic communications (e.g. the controller sending commands to an actuator every 200 µs)
- Aperiodic deterministic communications (e.g. fault indication from a device)
- Non-deterministic communications, with no time-criticality, or
- Some combination of the above (mixed traffic)

A critical service-dependent factor is the survival time; how long a break or how many lost packets does the URLLC application tolerate while still being able to perform normally. This depends on the application. A related factor is the possible consequence of exceeding survival time: what type of malfunction or anomaly is caused.

Instead of a single URLLC service requirement that could be used as input for transport analysis, requirements are service-specific. For transport, the environment and site where the services are delivered is of special importance. The case of a mining operation is different from URLLC service on an enterprise campus site.

In addition to the service requirements and sites, for transport, much depends on the selected network deployment architecture; a classical base station or disaggregated RAN with cloud technologies. Logical interfaces and the required transport connectivity depend on this too.

In many cases, a transport network is common to various applications and services and transport is required to be kept common for cost-saving reasons whenever technically achievable. When URLLC services are combined with other traffic flows, the common transport network has to be capable of supporting all traffic types without compromising QoS requirements for any of the flows. The most demanding service requirements typically originate from URLLC applications.

Network slicing, including related transport slicing, can serve as an important building block for URLLC services, as there may be a specific slice (or slices) supporting URLLC requirements (using slice type 2) and other slices for other services.

5.4.2 Reliability

In Chapter 4, reliability of transport was discussed with the outcome that transport links need to be considered as part of the system reliability analysis as they may contribute significantly to the estimated overall reliability. Some of the 5G system enhancements for reliability, like PDCP duplication, or 5G system operating as a bridge, can also cover transport links and increase reliability for transport. Redundant links and nodes are likely needed in transport as part of the system or as separate transport functionality; for example with packet duplication over disjoint paths. The need for this all depends on the estimated failure rates and on the targeted estimate for reliability figure.

A simple model for reliability on the air interface considers that bit error and packet loss rates can be compensated (at least to some extent) by retransmissions and multipath/multi-connectivity-based packet duplication. For transport links, the residual bit error rate in well-performing links is negligible and congestion-induced packet loss can be controlled by QoS and traffic engineering methods. What remains then is the probability of link and node failures, which can be dealt with by packet duplication over multiple links (multipath).

To get the ideal reliability gain with multipath, the two or more paths used should not suffer from the same failures, at least not at the same time. Failures on different paths should not correlate. With the air interface, the simple model in Chapter 4 assumed ideally that the first transmission and the retransmission do not correlate and also that the transmissions via one air interface path and another air interface path do not correlate. In the real world this is unlikely to always hold for the air interface.

For transport, using two fibre strands alone does not provide much resiliency if the same cable is used. To minimize the correlation, fibres on separate cable ducts can be used. This is generally very costly for wide-area transport, but it may be much more feasible in a specific smaller area, which is the case with URLLC services, at least for cases like factory plants, harbours, mines, etc. Ring topology may solve some of the cases, too. Another option is to use separate media, meaning a combination of optics and wireless. Some alternatives are shown in Figure 5.18.

Figure 5.18 depicts three cases where gNB connects for the URLLC services to a remote multilayer site switch, and via that further to the core network functions and an application server. In each case there are duplicated links but with different implementation.

The uppermost case illustrates high correlation between links, since the individual fibre strands used for the optical links 1 and 2 both fail simultaneously if the cable gets cut. In the middle case, two different cables (marked A and B) use different physical routes for the two links and the situation improves. The lowermost case shows a combination of optical and wireless links, where failures can be expected to be uncorrelated for many of the possible fault scenarios.

For the duplication methods, in addition to the PDCP duplication and GTP-U duplication in the 5G system, it is possible to have duplication at the application layer over the whole 5G system, with the 5G system operating as a bridge.

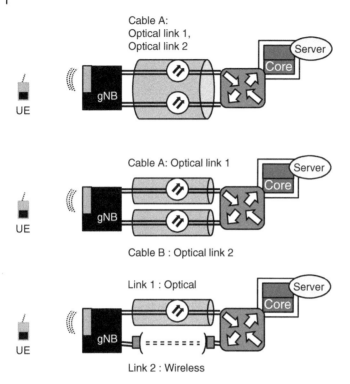

Figure 5.18 Disjoint paths.

At the networking layers, technologies for packet duplication are, for example, in IEEE the Frame Replication and Elimination for Reliability (FRER) (see [17]) developed in the TSN family of standards, and then IETF DetNet (see [18]) for the DetNet architecture, covered further in Ch 8.

5.4.3 Latency

The other part of the URLLC, low latency, is heavily impacted by the transport network layer, due to physical distance and related propagation delay. The primary way to mitigate this is to bring the core network function and the application servers close enough to the cell site, as in Figure 5.19.

The example in Figure 5.19 shows a case where the URLLC service is delivered by a local UPF located in a far-edge data centre, that is, located perhaps within a few kilometres from the cell sites and focusing on serving the URLLC applications. MBB services rely on a regional/core data centre, with more servers and more computing power available.

In the example, URLLC could use a network slice where in the transport domain the URLLC PDU sessions are routed to the local CU-U (user plane function), from there to the local UPF and further to the local server, for minimized latency and minimized amount of inter-site transport links. The CU-C network function is supported by the regional/core data centre, together with the core network control plane functions (AMF and SMF).

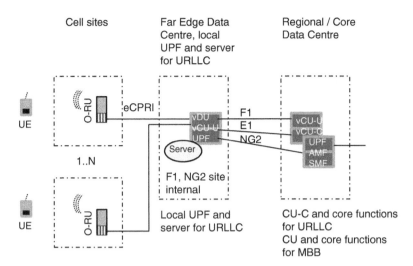

Figure 5.19 Local UPF and server.

MBB services are on a different network slice, and network functions are supported by the regional or core data centre. Variations in location of network functions exist both on the RAN side (CU, DU functions and, further, in separation of CU-U and CU-C functions) and then on the core network functions, so 5G architecture flexibility becomes visible there too.

Scheduling-related mechanisms in transport layers for guaranteed maximum latency point to traffic engineering-type approaches and IEEE TSN technologies.

5.5 Industry Solutions and Private 5G

5.5.1 Introduction to Private 5G Networking

Private 5G networks are often brought up in the context of the so-called Fourth Industrial Revolution, also known as Industry 4.0. Although Industry 4.0 is very much focused on the manufacturing industry, and while 5G technology certainly has a major role to play in enhanced digitalization and automation in manufacturing, private 5G networks are expected to play a significant role in all industries.

Private networks help solve even basic wireless connectivity in industries like agriculture, mining, oil and gas production and power generation, due to missing public network coverage and the infeasibility of WiFi systems. Private network markets also cover public safety and railway communication sectors, which require mission-critical service capability with often almost nationwide coverage. Here the focus is on industrial use cases with significantly smaller coverage, such as an industry campus, port or underground mine.

Discrete and process manufacturing industries such as chemical, petroleum, paper, food, car and electronic industries benefit from efficient and reliable private wireless 5G communication. Private 5G networks will also enhance service industries like logistics and transport. Therefore, for example, ports and airports are common within the early adopters of private 5G networks. Also, other enterprises in services and the entertainment sector are closely following 5G developments in order to maintain their competitiveness and efficiency. Public sector services typically must use the latest technology for improved cost efficiency, and therefore 5G is becoming crucial in hospitals and health care.

Common industry use cases for high-performance wireless communications include:

- High-quality real-time video in the uplink direction for tele-remote machine control and Artificial Intelligence (AI) applications such as quality control.
- Massive wireless video monitoring in large industrial campus areas.
- Low-latency communication for machine control and communication between collaborative autonomous robots.
- Ethernet connectivity for non-IP industrial protocols.
- High-performance mobility for automated guided vehicles, indoors and outdoors.

When various performance requirements are combined, it is evident that WiFi is not adequate technology. For example, large industry sites such as ports need connectivity in areas covering tens of square kilometres even. While 4G is excellent mobile technology for indoor and wide-area outdoor environments, it lacks capability to support extreme broadband and below 10 ms low-latency communication over the radio. 5G technology is designed to support a new spectrum enabling high-capacity communication, as well as an optimized 5G radio interface for ultra-low latency even down to 1 ms.

In summary, private 5G will be crucial in a wide range of industries, in addition to the traditional manufacturing factories. On the other hand, private 5G networks will not replace the need for mobile network operators' public communication services, which provide nationwide and global communications for enterprises and industries. Neither will private 5G replace existing WiFi use cases in enterprises such as WiFi laptops with office applications.

5.5.2 3GPP Features Supporting Private 5G Use Cases

5.5.2.1 Private 5G Deployment Options

3GPP has considered the requirements of a vertical domain from Release 16 onwards by specifying 5G architecture for private networks [1]. In 3GPP terminology, this is called-Non-Public Network (NPN). NPNs are divided into two flavors:

- Standalone NPN (SNPN), and
- Public Network Integrated NPN (PNI-NPN)

A standalone NPN is a fully isolated complete 5G system, which does not interact with any mobile network operator's 5G network (see Figure 5.20). Unlike a standalone NPN, a public network integrated NPN relies at least partly on 5G system functions on a mobile operator's infrastructure.

A public network integrated NPN can be a network slice allocated by an operator for the NPN or it can be a dedicated DNN provided by the operator (see Figure 5.21). In an operator-provided PNI-NPN, the UE has a subscription for the operator PLMN. If the PNI-NPN should be limited to a certain geographical location, then a Closed Access Group (CAG) can limit the access to selected cells or alternatively the PNI-NPN subscription is limited to certain tracking area(s).

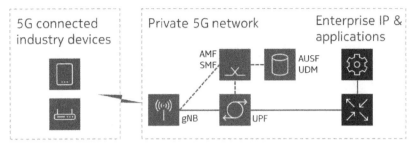

Figure 5.20 Standalone non-public network – simplified architecture.

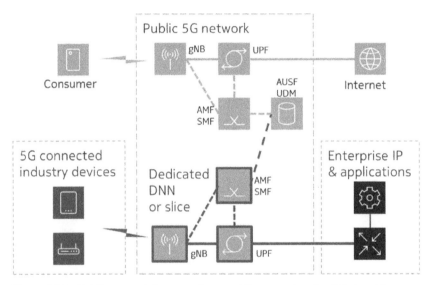

Figure 5.21 Public network integrated non-public network – simplified architecture.

The UE may prefer to connect to both NPN and public network services simultaneously. In that case, depending on whether the UE is in a public network coverage area or in an NPN coverage area, the UE can connect to other network services via N3IWF (Non-3GPP Interworking Function). Figure 5.22 shows an example of a UE, which is registered to an SNPN, registering to the PLMN via N3IWF. In this case, the SNPN has the role of 'untrusted non-3GPP access' and the UE establishes an IPsec connection to N3IWF via the SNPN user plane service. Control and user plane communication with AMF and UPF in the PLMN core is done over the IPsec connection.

The NPN can rely partly on the operator infrastructure, also based on RAN sharing. NG-RAN (gNodeBs) can provide connections to both public network and NPN 5G cores (see Figure 5.23).

5.5.2.2 Slicing in Private 5G Networks

5G slicing enables the establishment of isolated end-to-end logical networks, which share the same physical network infrastructure. Enterprises' requirements for slicing vary, and different properties are emphasized depending on the use cases. The slices can provide the following isolation properties:

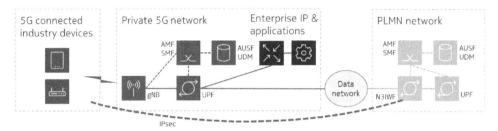

Figure 5.22 UE registered to standalone NPN and to PLMN via N3IWF – simplified architecture.

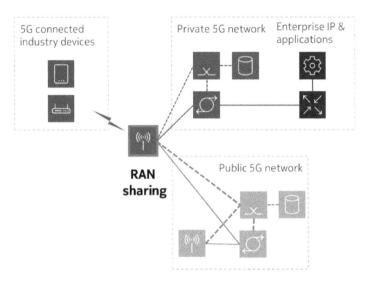

Figure 5.23 RAN sharing between public and non-public networks.

- Performance and QoS
- Security and privacy
- Fault isolation
- Management

In private networks, typical reasons for isolation between different enterprise applications or user groups are based on different QoS requirements and security isolation. For example, production applications and their users may be required to be separated for operations and maintenance networks. This separation in the 5G network can be implemented with slicing.

Slices in public networks serving enterprises are potential solutions if the public service provider has adequate coverage and capacity available. However, missing coverage is often a major problem in many industrial plants and locations. Mines, oil and gas and energy production sites are commonly far outside commercial coverage areas. Further, manufacturing factories and logistics warehouses often have poor indoor coverage provided by public operators. Even areas with good public coverage – such as airports or ports – may not have adequate guaranteed capacity for the industry use cases. Therefore, a private network may be the only option to utilize 5G slicing technology in industry plants and campuses.

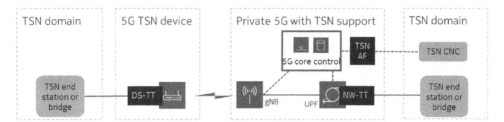

Figure 5.24 Private 5G as a logical TSN bridge.

5.5.3 URLLC and TSC in Private 5G

URLLC is one of the key E2E services supported by 5G slicing. URLLC requires overall E2E architecture support and especially many optimizations improving reliability and low latency in the 5G air interface. Low latency in radio is enabled with 3GPP standard capabilities such as short transmission duration (e.g. mini-slots), reduced processing in UE and gNB including faster HARQ timeline and grant-free uplink transmission. Low latency means also that physical distances must be reduced for user plane traffic. Especially in large enterprises, network design should bring the 5G core UPF functions as well as application servers close to the gNBs.

Reliability can be achieved with different techniques enabled by the 5G system. Packet duplication can be performed in the radio interface based on PDCP PDU duplication. Redundant data transmission can be done also on the higher layer using two PDU sessions with dual connectivity and redundant UPF nodes. The protocol using duplicate data paths can, for example, be FRER as specified by the IEEE.

Time-Sensitive Communication (TSC) relies on URLLC services in combination with features which allow the 5G system to act as a 'logical' Time-Sensitive Networking (TSN) bridge (see Figure 5.24). The 5G TSN bridge connects to the rest of the TSN with device-side TSN translator (DS-TT) and network-side TSN translator (NW-TT) functions. The 5G TSN system (i.e. DS-TT, UE, gNB, UPF and NW-TT) must be synchronized to the 5G grand master clock. The translator functions support synchronization with the TSN time domain according to IEEE TSN specifications.

URLLC and TSC services with deterministic traffic having bounded latency and jitter requirements must also have QoS mechanisms in backhaul and midhaul transmission networks. Technically, the simplest solution is to over-dimension transmission and use, for example, 10 or 25 Gbps Ethernet links with a small number of hops. If the load in the transmission links can be high, then E2E deterministic and low-latency communication requires QoS scheduling and queueing in transmission interfaces and elements. Backhaul and midhaul could be based on a dedicated TSN solution serving backhaul and midhaul connections. Alternatively, IETF-based deterministic networking (DetNet) for IP and MPLS with bounded latency support can be used.

5.6 Smart Cities

The role of cities keeps growing. Many factors are contributing to this evolution, including population migration to cities, higher specialization and innovation among dense populations and critical mass for new service paradigms based on digitalization. The global city challenge is how to make life better for citizens and how to meet the United Nations Sustainable Design Goals (UN SDGs). At the same time, most energy is consumed in cites. Cities need to be sustainable in all actions, but almost all areas need improvement. One big enabler for improvement is digitalization using IoT sensors, connectivity and machine learning, which can be called a new-generation digital ecosystem or digital city backbone.

Cities are or will be covered by Communication Service Providers' (CSPs') normal nationwide 5G mobile (macro) networks. This already enables many smart city applications and 5G gives many more features than previous mobile network releases. Quite soon, uplink traffic will increase dramatically – driven by high-capacity sensors like video cameras and lidars. These are needed for the many use cases in transportation, public safety, energy and people flow-related services. High-uplink traffic drives 5G mmW small cells to be implemented.

Small-cell city networks differ from macros, technically and in a business sense. It will be costly to build many small cells and specifically costly if own separate BH/FH for all operators is built into the city infrastructure.

Digitalized services, machine learning and data processing will be done in different locations based on the processing capacity and delay requirements, and partly to avoid privacy violations. Some of the processing may already be done in sensors like smart cameras, some in different edges and the rest in clouds. This will also impact on BH/FH dimensioning.

5.6.1 Needs of Cities

Our society and cities face great challenges to improve safety, energy efficiency, air quality, effectiveness of transportation and quality of living. The importance and role of digitalization is crucial in enabling cities' energy transition towards climate neutrality. A city has many systems and subsystems that need to be seamlessly interlinked and connected in a holistic and systemic way. Digitalization supports a city innovation ecosystem, to protect the environment, support the economy and promote the well-being of society.

City leaders are becoming increasingly aware of the need to design the fast transformation of infrastructures, processes and landscape in a holistic, multi-disciplinary and sustainable manner. A successful city needs to continuously develop its attractiveness and efficiency, while navigating through the complexities of urban digitalization.

To manage the uncertainties and fast innovation on the smart city application layer, cities need to secure availability of the underlying infrastructure at the right spots, especially electricity and communications. If not well planned, these infrastructures may create harmful timing bottlenecks due to heavy investments and potential legal hurdles.

City management ultimately needs a full sensor-based visibility of all necessary parts of city operations, as well as real-time control allowing optimization. A city is a natural unit of design with its geographic, demographic and legal boundaries. Many cities already have an integrated and holistic approach to city design but one important component – telecommunications – is still largely ignored. Especially radio base stations, which are the naturally visible part of the telecommunication network, and the needed BH/FH often emerge as an afterthought in the city landscape when developed incrementally by competing and uncoordinated mobile operators. At the same time cities don't like mobile operators digging fibres and opening roads continuously, and they don't like that there are antennas everywhere in the city landscape.

Cities' needs for connectivity are slightly different than CSPs' primary business offering incudes. Smart city digitalization and sustainability need a lot of information from everywhere in the city area. A lot of sensors need to be connected, partly through wireless networks and partly through fixed networks. Data from sensors, as well as other data from the city, should easily be shared with those who are creating and providing new sustainable digital services inside the city.

Co-operation between city and mobile operator is needed anyway, albeit current mobile operators and smart cities may have different needs for the network. Cities use the network as a platform for their digital ecosystem and to increase the quality of living for their citizens and meet the UN SDGs. It is also possible to add mobile operators' terminal data to the city data-sharing markets, so the mobile operators have a possibility to monetize their crowdsourced data.

5.6.2 Possible Solutions

Higher-frequency air interfaces are needed as soon as smart city data traffic increases because of surveillance cameras, 3D streaming and other data-hungry services which mainly use the uplink direction. High frequency enables high data capacity, but the high-frequency signals span only short distances, which means that a lot of small cells are needed. To deploy a small cell every 50 m is a huge challenge for mobile operators. One way to solve the deployment is to start using light poles for 5G small cells. Light poles are located normally 40–50 m distance from each other and are often owned by the city. To combine the smart city needs and the mobile operator deployment challenge by using (city-owned) light poles as a site for 5G radios sounds logical. The city would need to support the light poles with constant electricity and data networks. The light poles will transform to smart poles, having many sensors and 5G base stations integrated. A smart pole network will be used with the city and mobile operators in sharing mode. Sharing of course also reduces the cost challenge.

The new digital ecosystem of smart cities requires open interfaces and data access for different stakeholders and service providers. Figure 5.25 describes new smart city network layers, which help cities to create new sustainable data-sharing ecosystems.

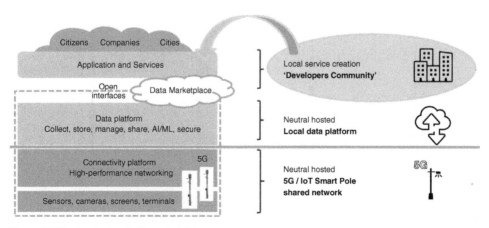

Figure 5.25 Example of city network layers.

On the bottom there is shared neutral-hosted connectivity and sensor layers located mainly in smart poles. The city or other neutral party owns the smart poles and fibre BH/FH. Mobile operators share the 5G small-cell network by leasing the capacity from antennas or the place for their own 5G base stations.

On top of the connectivity layer there is the city's or other local neutral player's data platform and data marketplace. A locally and neutrally orchestrated data marketplace enables secure data sharing and smart contracts for sharing based on city-level rules and privacy aspects. Data comes from sensors, but it can also be data from mobile terminals or other data from the city area.

This platform makes it possible to locally create new city services based on the data from the marketplace. Service creation may also be done by the local community of developers. This can be called as new real smart city digital ecosystem. This kind of digital city backbone is presented in the LuxTurrim5G Ecosystem in Finland [19]. Cities worldwide have been interested in building these kind of layers, together with mobile operators and other actors, but so far the business models are in the infant phase.

Different smart pole alternatives are presented in Figure 5.26. The most flexible and fully featured smart poles are connected by fibre and have IoT sensors and shared 5G mmWave antennas (leftmost pole in Figure 5.26). Quite often in the city area, fibre is not available near the pole and it may be costly and time consuming to dig and connect. Other possibilities are to use E-band or in future D-band mmWave radios or IAB. If the 5G mmWave antennas are not needed in the pole, the pole can be connected by 5G CPE (rightmost pole in Figure 5.26).

One scenario is that in the beginning, 5G mmWave access is used mostly as Fixed Wireless Access (FWA) to connect those smart poles to a network which don't have a 5G mmW antenna or fibre. In those cases, the connectivity comes through the nearest 5G mmWave antenna, mostly inserted into another smart pole (rightmost pole in Figure 5.26). And then step-by-step when lower-cost 3GPP Release 18 RedCap CPEs are available, the connection to sensors near the poles and, for example, in moving cars will go through the nearest 5G mmWave base station.

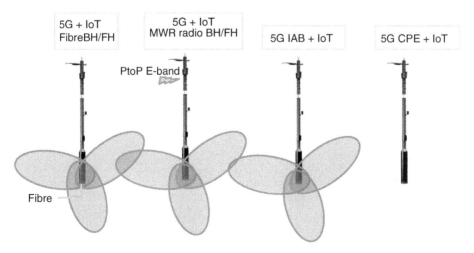

Figure 5.26 Smart pole alternatives.

The frequency may be private, even city owned if country regulations allow, but it may be some type of tower company taking the role of neutral host, or even a mobile operator. Step-by-step all the mobile operators in the area may share the network at some level, but using their own frequency bands. The IoT devices in the poles will use the same transport and at the point of local breakout, the IoT data and 5G data can be integrated and prepared for the data marketplace.

Figure 5.27 shows the possible sharing alternatives. There is also pressure to have the far edge – even inside the smart pole – driven by data processing, especially for URLLC or data with personal information (video streams).

5.6.3 New Business Models

There are many different alternatives for business models in smart city networks. Quite often, many companies are needed with different roles. In Figure 5.28, different main actors are presented. The actors may be different depending on the sharing models presented in Figure 5.29.

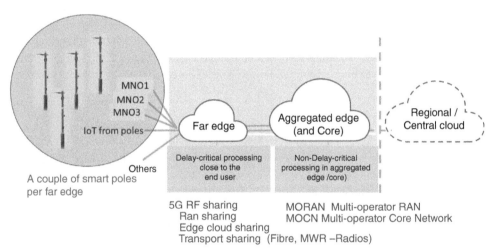

Figure 5.27 Shared city network.

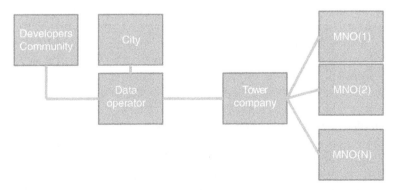

Figure 5.28 Example of actors in digital city backbone network.

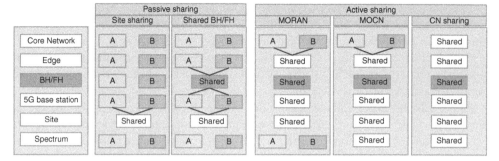

Figure 5.29 Infrastructure sharing.

Passive sharing with poles and sensors may be owned by the city or by an advanced tower company; the BH/FH or at least fibre ducts may be included. The tower company may also have a local data centre and in the ultimate case also 5G mmWave base stations with their own private frequency and some part of the edge computing. Mobile operators lease the site for the poles from their own 5G mmWave radios and antennas, or share the 5G mmWave base station and antenna with others. The data platform and data marketplace may be owned and operated by the city or a separate data operator, or by the tower company. Of course, many other alternatives may exist in different countries based on their governance models.

5.6.4 Implications for BH/FH

Small-cell city networks have a dense BH/FH network with moderate short hops. This will be quite costly to build if it needs to be done in an already built environment. Digging the fibre ducts under the streets to every smart pole is expensive and takes time. Here the alternatives are earlier presented mmWave links or IAB. For new city areas, the fibre ducts should be installed at the same time as streets are laid. This is the opportunity for cities to start building the new digital backbone. It will be a valuable asset for the future, as one day anyway the poles will need to be connected and the city can foster mobile operators to use shared BH/FH and also monetize the investment. The city may also install their own fibres to the ducts to control the IoT data from sensors and lease the BH/FH capacity to mobile operators.

There are many ways to share the network. It may be passive or active sharing. In passive sharing, only the site (here the pole) is shared and the BH/FH is not. Or maybe also the BH/FH is shared. In active sharing there are three possibilities: Multi-Operator RAN (MORAN), Multi-Operator Core Network (MOCN) and Core Network (CN) sharing (see Figure 5.29).

In MORAN, case operators have their own frequency and core network; other parts (including BH/FH) are shared. In MOCN, the frequency is also shared. In CN, the whole network is shared.

In city networks all these are possible but maybe most relevant is MORAN, or in some cases passive with BH/FH sharing. This depends on the city network-related actors and their business models. Anyway, in most cases the BH/FH is shared.

City networks can also be built so that there is only one operator's 5G base station in one pole. One pole is only for one operator and the IoT, the other pole is for another operator and the IoT, and so on. In this case also the BH/FH is preferably shared, to minimize the fibres. Normally, not all the poles need 5G mmWave antennas; coverage is reached with fewer of these, but all the poles need at least some sensors connected to the network, at least for smart lighting. In Figure 5.27, different pole connectivities are presented.

There are many technical things to be taken care of in city network BH/FH. More detailed descriptions are presented in this book: IAB in Section 5.2; URLLC in Section 5.3; private networks in Section 5.5; optical solutions in Chapter 6; and wireless BH/FH in Chapter 7. Many operators may use slicing also in the shared network; slicing details are presented in Section 5.1.

City networks have their own specialities, some of those impacting the normal BH/FH technical solutions. For example, in a shared network the normal traffic load gets closer to peak capacity than in one operator network. IoT data from the sensor in the pole needs to be added to the capacity planning. If IAB is used, the IoT data needs to be connected through an extra CPE in the pole while IAB is only carrying the 5G IAB data, not the separate IoT data. In shared BH/FH the resilience and privacy requirements may vary, especially for IoT data which is not based on 3GPP.

Edge functionality moves the data processing from the core towards the network nodes. It will reduce the need for BH/FH at some level. If the UPF (local breakout) from 5G is in the edge, this makes it possible to combine the 5G and IoT information already in the edge. Private data, like data from video cameras, may be forced to locate to the far edge – as near as possible to the cameras. This drives part of the edge functionality inside the smart poles and also affects the BH/FH traffic.

On the other hand, when building long-lasting city networks, it will be good to prepare for 6G BH/FH traffic, since those networks will be needed during the lifetime of a city infrastructure built for 5G.

References

1 3GPP TS 23.501 System architecture for the 5G system (5GS).
2 3GPP TS 38.473 NG-RAN; F1 Application protocol (F1AP).
3 MEF 84 Subscriber Network Slice Service and Attributes.
4 3GPP TS 38.300 NR; NR and NG-RAN Overall description.
5 3GPP TS 38.401 NG-RAN; Architecture description.
6 3GPP TS 33.501 Security architecture and procedures for 5G system.
7 3GPP TS 38.425 NG-RAN; NR user plane protocol.
8 3GPP TR 38.811 Study on New Radio (NR) to support non-terrestrial networks.
9 3GPP TR 23.737 Study on architecture aspects for using satellite access in 5G.
10 3GPP TR 38.321 Solutions for NR to support Non-terrestrial networks (NTN).
11 https://1.ieee802.org/tsn/
12 https://datatracker.ietf.org/wg/detnet/about/
13 3GPP TS22.261 Service requirements for the 5G system.
14 3GPP TS 22.104 Service requirements for cyber-physical control applications in vertical domains.
15 3GPP TS 22.186 5G Service requirements for enhanced V2X scenarios.

16 3GPP TS 22.289 Mobile communication system for railways.

17 IEEE 802.1CB Frame replication and elimination for reliability (FRER) [802.1CB].

18 IETF RFC 8655 Deterministic networking architecture.

19 https://luxturrim5g.com.

6

Fibre Backhaul and Fronthaul

Pascal Dom, Lieven Levrau, Derrick Remedios and Juha Salmelin

This chapter describes fibre-based backhaul/fronthaul (BH/FH). Optical transmission and fibre are essential parts of the whole Transport Network (TN). In order to complement optical transmission, packet networking with IP and Ethernet is needed. These technologies are mostly covered in Chapter 8 except for TSN (Time Sensitive Networking) for fronthaul, which is often coupled with optical transport, and which is also discussed here.

Chapter 7 presents wireless BH/FH, which is also part of the TN, but mainly used for last-mile connectivity in case fibre is not feasible.

6.1   5G Backhaul/Fronthaul Transport Network Requirements

The flexible decomposition of 5G baseband functionality into Radio Unit (RU), Distributed Unit (DU) and Central Unit (CU) functions has created fronthaul, midhaul and backhaul segments, and driven a need for a flexible transport infrastructure to support the multiple protocols that could be present in the network – especially for the fronthaul. Each segment has requirements related to capacity, latency tolerance, aggregation capability, synchronization and supported port rates, as summarized in Figure 6.1. The challenge for a network operator is to provide a TN that can support the requirements and flexibility, whilst at the same time ensuring that it is not over-engineered for the applications intended to be delivered, in order to maximize the return on investment for the solution.

Other considerations for the 5G BH/FH TN discussed in this section are related to capacity, latency, synchronization, availability, network slicing and programmable network with SDN (Software Defined Networking) and OAM (Operations, Administration and Management) capabilities.

Please refer to Chapter 4 for a discussion on the requirements for 5G backhaul and fronthaul transport.

6.1.1   Capacity Challenge

The capacity requirements for fronthaul, midhaul and backhaul are dependent on the radio technology used (sub-6 GHz vs mmW), symmetrical or asymmetrical functional splits and number of radios at a particular site. Generally, multiplexing gain increases with

5G Backhaul and Fronthaul, First Edition. Edited by Esa Markus Metsälä and Juha T. T. Salmelin.
© 2023 John Wiley & Sons Ltd. Published 2023 by John Wiley & Sons Ltd.

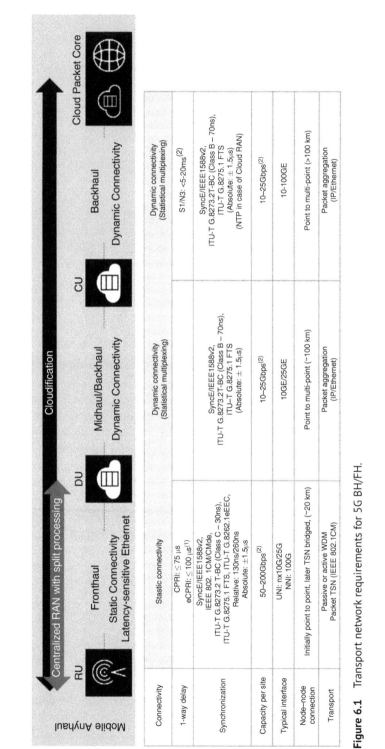

Figure 6.1 Transport network requirements for 5G BH/FH.

	Static connectivity	Dynamic connectivity (Statistical multiplexing)	Dynamic connectivity (Statistical multiplexing)
Connectivity			
1-way delay	CPRI: ≤ 75 μS eCPRI: ≤ 100 μS[1]	S1/N3: <5–20ms[2]	
Synchronization	SyncE/IEEE1588v2, IEEE 802. 1CM/CMde, ITU-T G.8273.2 T-BC (Class C – 30ns), ITU-T G.8275.1 FTS, ITU-T G.8262.1eEEC, Relative: 130ns/260ns Absolute: ± 1.5μs	SyncE/IEEE1588v2, ITU-T G.8273.2T-BC (Class B – 70ns), ITU-T G.8275.1 FTS (Absolute: ± 1.5μs)	SyncE/IEEE1588v2, ITU-T G.8273.2T-BC (Class B – 70ns), ITU-T G.8275.1 FTS (Absolute: ± 1.5μs) (NTP in case of Cloud RAN)
Capacity per site	50–200Gbps[2]	10–25Gbps[2]	10–25Gbps[2]
Typical interface	UNI: nx10G/25G NNI: 100G	10GE/25GE	10-100GE
Node–node connection	Initially point to point, later TSN bridged, (~20 km)	Point to multi-point (~100 km)	Point to multi-point (>100 km)
Transport	Passive or active WDM Packet TSN (IEEE 802.1CM)	Packet aggregation (IP/Ethernet)	Packet aggregation (IP/Ethernet)

functional split. The lowest-layer split protocols on legacy fronthaul (3GPP Option 8: CPRI/OBSAI) transport the on-air information in a continuous stream, without the ability to statistically multiplex traffic from multiple radios. Higher-layer split and backhaul streams are similar in nature to the end user traffic characteristics and, with low time and spatial correlation between physical users and radios, allow greater multiplexing gains over the TN. 5G fronthaul and packet networking with eCPRI/O-RAN protocol stack lies between these extremes.

6.1.2 Latency Challenge

There are two factors that impact the latency limitation for a 5G network (see further details in Chapter 4):

i) eCPRI HARQ (Hybrid Automatic Repeat reQuest) loop latency – this limits the round-trip delay of the TN to 0.3 ms round-trip time up to a far edge location where the HARQ termination is in the DU. Hence the maximum distance (speed of light in glass) at 300 μs is 30 km TN for a low-layer split with fronthaul.

ii) Application latency – this is the round-trip time from the user through the 5G baseband processing, 5G Core Network (CN) and end application. For ultra-low latency applications, this will result in the baseband processing (DU/CU and local CN on a Multi-access Edge Compute, MEC) being co-located at the cell site with the end user.

Next Generation Mobile Networks (NGMN) has provided recommendations and considerations that need to be made in determining where to place the 5G DU/CU/UPF functions to meet the latency requirements of baseband processing and application (Figure 6.2) [1]. The first row provides a general view of scale – with a factor of 10 between each data centre tier. If a User Plane Function (UPF) is located in a Tier 1 site, then 1000 Tier 1 sites will have their own version of the UPF. If the UPF is moved to a Tier 2 site, then only 100 versions are required. This level of pooling can save the operator considerable capital expenditure/operating expenses (CAPEX/OPEX) in running the UPF, as well as potentially gaining benefits from traffic behaviours (e.g. mobile tidal effects when populations move from suburbs to city centres during the working day). However, the UPF can only be centralized so far as the latency budget is still met, plus other factors such as transport capability. It can be seen that 1 ms uRLLC applications would co-locate all Baseband Unit (BBU) processing and UPF at the cell site. However, 10 ms low-latency eMBB applications have the option of locating the DU at the cell site or Tier 1 sites with the CU/UPF at Tier 2

Sites	Cell Sites	Tier 1 / Far Edge	Tier 2 / Edge	Tier 3 / Core
Number	100,000's	1000's	100	10
Transport latency (one way)	0	0.6ms	1.2ms	4.2ms
RTT uRLLC	1ms	2.2ms	3.4ms	9.4ms
RTT eMBB	8ms	9.2ms	10.4ms	16.4ms

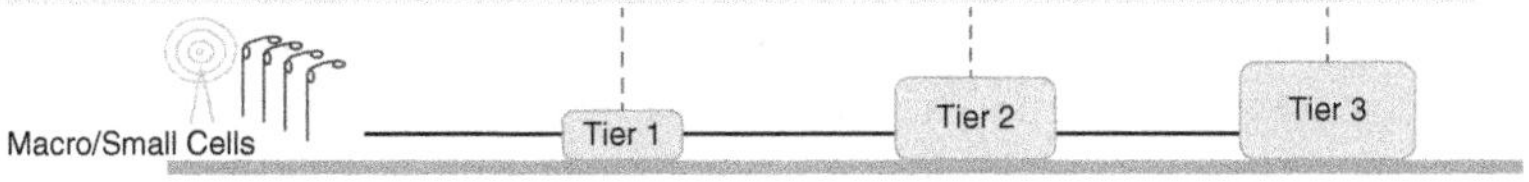

Macro/Small Cells — Tier 1 — Tier 2 — Tier 3

Figure 6.2 Data centre latency requirements [1].

(edge) or Tier 3 (core) sites. This provides a reasonable trade-off between cost and performance – and also adds a degree of freedom in the choice of providers for the access segment of the connection. As the location of the UPF is dependent on the application, the provider should have a flexible option to (re-)locate the UPF to the most economical point; this can be achieved using SDN in both RAN and TN.

6.1.3 Synchronization Challenge

The synchronization requirements (frequency and phase/time) of a 5G RAN can be split into two groups:

1) 3GPP defines Time Alignment Errors (TAEs) for a 5G antenna operating in the Time-Division Duplexing (TDD) spectrum as 3 μs, which can be supported by a TN that assures a maximum absolute time error (max|TE|) of 1.5 μs from the grandmaster (time) clock. This will set the challenge for a TN used for backhaul or midhaul.
2) 3GPP also defines the TAEs inside a (fronthaul) cluster of base stations, depending on the spectrum used for carrier aggregation (frequency range intra- or inter- (non-)contin-uous bands) with categories of relative TE: A = 130 ns, B = 260 ns, C = 3 μs. This will set the challenge for a TN in fronthaul.

6.1.4 Availability Challenge

As with latency, the desired applications drive the availability challenge. Application or communication service availability is the up-time of a service from the user of a 5G end-to-end network (RAN, TN or CN). For ultra-high reliability applications, this can be expressed in a number of nines. In 3GPP, Annex F describes the relation of reliability and communi-cation service availability [2]. TN availability is part of the reliability as defined by 3GPP. Refer to Chapter 4 for a more detailed discussion.

6.1.4.1 Optical Protection
Optical 1 + 1 protection uses a scheme where the signal is bridged at the source simulta-neously between two diversely routed fibre paths to the destination. The receiver selects the better of the two arriving signals based on signal quality, or an external control signal. Signal degradation in the working will cause the receiver to switch to the protected channel. The simplicity of the technique is a trade-off with the extra capacity that is required for the protection path. Each working/protected path needs to be completely node disjoint and link disjoint. This protection mechanism can be deployed with a splitter and combiner.

This protection mechanism requires two diversely routed paths through the network. In most cases, the two fibres run through the same duct and consequently don't increase the availability, as typically the duct is severed – breaking both paths. An operator might choose to install in addition differently routed optical ducts, but that comes at a cost. Additionally, this protecting mechanism doesn't provide node resiliency.

Resiliency in the fronthaul network is envisioned to be deployed on a packet level, such as ITU-T G.8031 Ethernet linear protection switching [3] and ITU-T G.8032 Ethernet ring protec-tion switching [4] at Ethernet level, or MPLS-TP or IP/MPLS-based resiliency mechanisms.

6.1.4.2 Passive Optical Network Protection

A Passive Optical Network (PON) defines an alternative to a full duplex optical 1 + 1 protection. This alternative scheme is referred to as 'Type-B' PON protection. It leverages the capability of a TDM-PON solution to full duplex protection of the 'resources' that are shared between multiple cell sites, with a simplex solution for single-site resources. Last (drop) fibres and Optical Network Termination (ONT) are dedicated to a single site and can be simplex, where Optical Line Termination (OLT) and feeder fibres are shared resources that need full duplication (1:1). For passive feeder fibres, 1:1 means fully diverse routed cables; for OLT this means redundant (geographically separated) instances. The above-described protection is specified as 'PON Type-B' and can achieve a 50 ms switch over time and short drop cables with five-nine TN availability.

6.1.5 Software-Controlled Networking for Slicing Challenge

After the E2E bandwidth, latency and reliability requirements of services such as uRLLC, mMTC and eMBB are satisfied, traditional available Quality of Service (QoS) policies can be applied. Furthermore, to meet different types of user traffic and OAM requirements, adequate E2E network slicing must be provided; additionally, network slicing creates new business models for service providers and carriers while resolving the Service Level Agreement (SLA) requirements of various applications and users. This slicing technique requires agile programmability delivered by software controllers (SDN) as part of the OAM centre.

6.1.6 Programmability and OAM Challenges

Delivery of mobile transport services requires a number of operations to occur properly and at different levels in the service delivery model. Ethernet OAM is discussed further in this section.

To address various applications and requirements, Ethernet OAM is defined at several network layers. Each layer in the network uses a dedicated OAM protocol. The network provider, the operator and the end customer deploy their specific protocol to monitor network performance at their level of the hierarchy. These different OAM layers operate in parallel on the network.

In order to verify that the mobile transport service is operational, carrier-grade OAM technology must be supported, designed to operate over carrier Ethernet networks with both carrier-grade OAM and carrier-grade resiliency and availability. Each transport layer in the protocol stack should be supplemented with diagnostics specialized for the different levels in the service delivery model.

The Ethernet suite of OAM diagnostics provides management capabilities for installing, monitoring and troubleshooting Ethernet networks, which are especially relevant for 5G fronthaul in the Ethernet phase, based on the IEEE 802.1ag standard, which defines protocols and practices for OAM for paths through 802.1 bridges and local area networks. ITU-T Y.1731 complements this, with additional fault management as well as performance monitoring protocols in Ethernet transport networks, and MEF 35.1, 30.1 and 17 meet the requirements of MEF service OAM performance monitoring [5–9].

Table 6.1 Ethernet OAM.

Network layer	Protocol	Function	Description
Physical layer	DDM	Digital diagnostics	Monitors receive and transmit power, temperature and voltage levels
Link level	IEEE 802.3 link OAM	Discovery	Monitors communication between the devices on the link
		Critical events	Reports link failures and dying gasp
		Latching loopback	Enables link testing with measurement equipment
		Event notifications	Reports link-level code errors and frame errors
Service level	IEEE 802.1ag CFM and MEF 30.1	Connectivity loopback	Proactively monitors path between devices across a network
		Loopback	Discovers remote devices or localizes/isolatesfailures on demand
	ITU-T Y.1731 and MEF 35.1	Loss measurement	Supports frame-based measurement of frame loss and frame loss ratios
		Synthetic loss measurement	Supports synthetic message-basedmeasurement of frame loss and frame loss ratios
		Delay measurement	Supports measurement of frame delay

These standards are complementary to IEEE 802.1CM and are essential for carrier-grade Ethernet transport networks. For additional optical-layer diagnostics and monitoring, Digital Diagnostic Monitoring (DDM) should be considered. Ethernet OAM is defined atseveral network layers, as summarized in Table 6.1.

For in-band testing and verification, in-band packet-based OAM closely resembling customer packets is used to effectively test the customer's forwarding path, but these are distinguishable from customer packets so they are kept within the service provider's network and not forwarded to the customer.

The carrier-grade OAM Ethernet protocol suite should be supplemented with diagnostics specialized for the different levels in the service delivery model.

6.2 Transport Network Fibre Infrastructure

6.2.1 Availability of Fibre Connectivity

To satisfy the higher capacity and lower latency expectations of 5G wireless technology in the future, densification of cell sites will be a key requirement. High-band radios will have lower propagation ranges, resulting in a smaller intercell distance, leading to a higher cell-site density. Cellular operators will seek a cost-effective means to provide the required connectivity for these dense cell deployments, and while wireless backhaul solutions such as microwave and Integrated Access and Backhaul (IAB) in the mmW will have a place, fibre-based connectivity will remain the technologically preferred option due to its high capacity and low latency characteristics. Thus, economically delivering fibre connectivity to a dense cell-site edge is considered to be a foundational requirement of the rollout of 5G technology. Whilst the economics of delivering fibre connectivity in these situations is complex and dependent on specific deployment models and scenarios, a key facet of the solution involves the ability to share the transport infrastructure amongst many users. Sharing can be accomplished in many ways – from RAN sharing of the mobile infrastructure to sharing of the transport layer in terms of optical wavelengths (Wavelength Division Multiplexing, WDM) or multiplexing the signal of multiple users in time via a packet multiplexing process in switches as defined in the IEEE 802 series or Time Division Multiplexed Passive Optical Networks (TDM-PON).

The process of deploying multiple-strand fibre bundles in new ducts can be an expensive undertaking, with a long planning, regulatory and implementation cycle – so the selection of the routes becomes one of the most business-critical decisions for a fibre infrastructure owner. It is therefore important to maximize the use of existing infrastructure such as empty ducts, to avoid the necessity of digging and the associated costs. Consideration is given to mixed scenarios: laying cables in existing ducts where available and combining newly installed ducts and aerial cables where no ducts exist. Several street-level decisions are made: for example, minimizing street crossings, evaluation of road surface variations, avoiding conservation areas. The costs between various deployment routes are calculated and this may not necessarily result in the shortest (or lowest latency) path. The fibre network itself is a long-term investment, with a designed lifetime of at least 25 years and an operating lifetime typically much longer. The fibre infrastructure providers who finance the construction are required to amortize this expense over a finite timeframe as defined for a fixed asset class, and must obtain sufficient customers to lease fibre or wholesale this critical infrastructure.

The fibre infrastructure provider will typically have an opportunistic strategy of laying fibre when other major construction projects are underway and ducting tunnels are exposed, with particular interest in the ducts that pass by longer-term 'potential' customers such as multi-office tower blocks or 5G tower locations. From the entire bundle of fibre deployed, fibre strands can then be leased entirely (known as 'dark fibre') and dedicated to the customer for their services, or through an active multiplexing technology can be used to provide wholesale services for multiple tenants, via a third-party network infrastructure provider.

In the first case, the entire lease cost of the fibre strand is assigned to one customer on a monthly basis for, typically, a 3 to 5-year term. In the second case, the fibre cost is divided by the number of customers utilizing the multiplexed service. In both cases, an assumption needs to be made on the 'take rate' (the number of customers who will purchase the resource) for fibre or services, in order to calculate the resulting lease rate and ensure that the project will have a net positive Return on Investment (ROI).

6.2.2 Dedicated vs Shared Fibre Infrastructure

To further enhance the return on their fibre investments, operators will also target transport for business and enterprise customers – with the additional requirement of high availability of the transport layer with equipment redundancy and dual fibres on diversely routed paths.

The terms 'dedicated' and 'shared' fibre infrastructure relate to the ownership and use of the fibre. Dedicated fibre is deployed by a vertically integrated operator (such as a large Communications Service Provider, CSP who owns and deploys both mobile and transport infrastructures). The fibre, owned by the CSP, can be used when available with very low-cost optics at each end – providing the most cost-effective transport option. However, in many cases, fibre is not readily available and a multiplexing (WDM or packet) technology must be used over a leased dark fibre, for a cost-effective solution to be obtained.

Shared infrastructure is owned and deployed by a third-party fibre/network provider, with the aim of transporting multiple Mobile Network Operators (MNOs) who either lease the fibre strands (dark fibre) or lease a managed service.

Figure 6.3 shows radio site connectivity options for dedicated infrastructure (using dark fibre) and shared infrastructure (offered as a service by a network infrastructure provider). In the first case, with a single owner, all monitoring, control and management falls within one administrative domain, and failure segmentation is straightforward. In the second case, due to the fact that the network infrastructure provider needs to offer strict guarantees of link performance and availability – the SLA – additional functionality of demarcation (allowing tenant and infrastructure provider to agree on the interface performance) is required.

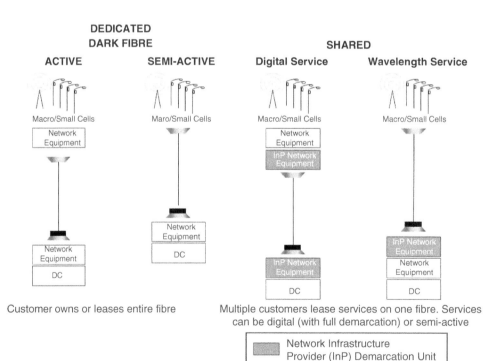

Figure 6.3 Connectivity options.

6.2.3   Dedicated Infrastructure

For a dark fibre solution, dedicated to a single MNO, three configurations are available: active, semi-active and passive. The active case uses the MNO's equipment to process (e.g. CPRI to RoE or eCPRI adaptation, see IEEE 1914.3 [10]) and aggregate the cell-site traffic onto higher-rate (100 Gbps) network interfaces.

Other functionality may also be included in the cell-site unit, such as the extension of the data centre spine/leaf fabric to the cell site to provide redundancy and high transport availability, as well as sync distribution. These are all configured and managed by the owner, the MNO. For the second and third cases, semi-active and passive, hardened, coloured optical transceivers are installed directly into the radio unit, and optically multiplexed in a passive WDM unit located nearby. The key benefit of these configurations is the simplicity and cost of the radio site, especially for sites with a low number of radio connections. The semi-active case differs from the passive case, with the addition of a monitoring unit at the central site to extract optical-layer performance and continuity metrics from the network. For a dedicated network with a single owner who owns both ends of the link, active, semi-active and passive configurations allow for performance monitoring and failure segmentation due to the single administrative domain covering both transport and mobile.

6.2.4   Shared Infrastructure

For a shared infrastructure with a network infrastructure provider offering BH/FH as a service connectivity, there are also two options available: digital and wavelength services.

A digital service is one which packet transports a bitstream that has previously been processed and aggregated by an MNO for their own radios, using an active demarcation unit at the cell site. If there are several MNOs present at the cell site, each would process and aggregate their own traffic and present the aggregated stream to the network infrastructure provider to transport, and these streams would be kept separate and isolated in the transport process (with hard isolation for the more restrictive fronthaul protocols, to ensure no interactions that could impact the latency of each stream; for other BH/FH protocols with more tolerant latency requirements, a 'soft-isolation' packet multiplexing technique could be used). The demarcation equipment needs to be hardened, support 25/50GE and even 100GE clients from the MNOs present at the site, and ensure that the 5G sync distribution requirements are met in a TN (fronthaul cluster or back and midhaul TN challenges).

A wavelength service is similar to the semi-active configuration of the dedicated use case but requires a demarcation unit at the central site to provide some level of network monitoring and demarcation. This allows the network infrastructure provider to provide a lower-cost service at the small expense of a reduction in network visibility. In conclusion, as described above, the introduction of flexibility in the mobile layer, together with diverse ownership models, fibre availability and support of multiple traffic protocols with disparate requirements, drives the need for a 'toolbox' of solutions to fully cover the 5G transport landscape.

6.3 New Builds vs Legacy Infrastructure

As described earlier, to meet the requirements of the latency challenge, the operator needs to decide where to place the various baseband functions (UPF) and CN (MEC) to serve his targeted application use cases (eMBB, URLLC and mMTC). In many cases, a suitable location (cabinet, building or data centre) is not available, and new infrastructure must be built, with adequate fibre connectivity to the cell sites and service locations.

An alternative option for fibre connectivity is to build an overlay on a brownfield access fibre infrastructure. Fibre to the Home (FTTH) fibre infrastructure, built to support broadband to the home and enterprise locations, is such a brownfield infrastructure. This FTTH infrastructure is, in many cases, based on point-to-multipoint passive splitters with a passive (not powered) remote cabinet site as shown in Figure 6.4.

The fibre route of an FTTH network is called an Optical Distribution Network (ODN) and spans from the central office (CO) to the endpoint (homes). The ODN consists of several sections of fibres joined with splices or connectors but provides a continuous optical path. Point-to-point topologies (P2P) provide dedicated 'home-run' fibres between the CO and the endpoint. Point-to-multipoint topologies (P2MP) provide a single 'feeder' fibre from the CO to a branching point and from there one individual dedicated 'drop' fibre is deployed to the endpoints. A similar topology to P2MP is a tap, chain or linear topology where multiple endpoints are on a single fibre trail towards the CO. A key characteristic of FTTH ODN is the use of a single fibre strand transport technology or bidirectional (BiDi) working on a single fibre strand.

For resilience in P2MP topologies there are three passive elements that can be protected within the fibre network:

1) The feeder facility (the ducts and fibres from the CO to the fibre branching point or splitter)
2) The passive optical splitter itself
3) The distribution facility (the duct and fibres from the branching point or splitter to the endpoint)

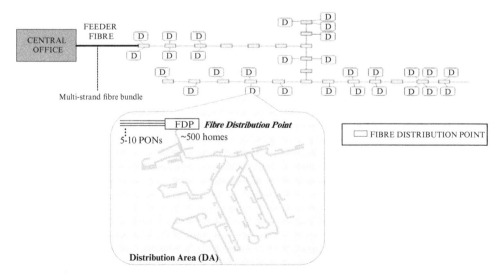

Figure 6.4 Transport network for FTTH.

Full redundancy, specified as Type-C PON protection, requires that all three topology elements need full duplex, which means the fibre is diversely routed. The advantage of P2MP topologies is having additional resilience options where only the feeder sections are diversely routed (e.g. via a ring architecture interconnecting the branching points), specified as Type-B PON protection.

We distinguish two versions of P2MP ODN for a PON, depending on the type of splitters used:

1) The splitter can be a passive optical power splitter that transfers feeder light into all branches. This requires a wavelength filter at the endpoint for selecting the spectral wavelength when used in a multiple channel system. On a BiDi P2MP ODN, multiple technologies can co-exist, with each technology using a different channel on a different wavelength pair. The endpoint selects its channel. Hence this ODN requires Wavelength Selection (WS) and is known as WS-ODN.

   WS-ODN has a typical optical budget of 28 dB that corresponds to about 20 km with a 32-way split, yet other budgets and split ratios can be deployed. The optical power splitters have an insertion loss that corresponds to around 3 dB for every two-way split. The vast majority of FTTH or FTTP-installed base brownfield cases around the globe are power splitters, hence WS-ODN. The sharing needs to address this WS-ODN. The different transport network technologies for WS-ODN are the TDM-PON described later.

2) The splitter can be a passive optical wavelength splitter (e.g. Arrayed Waveguide Gratings, AWG) that transfers a specific wavelength into each branch. On a BiDi P2MP ODN, multiple channels also use different wavelength pairs, yet each wavelength is routed to one branch. Hence, this ODN is known as Wavelength Routed (WR), or WR-ODN. WR-ODN has a smaller optical budget of between 20 and 24 dB that corresponds to about 20 km, as the AWG has an optimized insertion loss compared to the optical power splitter. For dedicated greenfield deployments of P2MP for fronthaul, a WR-ODN has been deployed.

6.4 Optical Transport Characteristics

6.4.1 Optical Fibre Attenuation

Optical fibre can support the transmission of light over long distances due to the low loss per unit distance. Low attenuation is obtained through the removal of impurities in the fibre itself – yielding a wavelength-dependent attenuation characteristic as shown in Figure 6.5, with a water absorption peak at 1400 nm.

The figure shows the attenuation spectrum for pure silica glass, where the intrinsic absorption is very low compared to other forms of loss. For this reason, optical communication systems are designed to operate in the region from 800 to 1600 nm. The units of attenuation of light in fibre are dB/km – showing the exponential relationship of loss (a 3 dB loss is equivalent to half the optical power). The water peak attenuation that was common in older fibre designs is shown, together with the flattened characteristic available with most modern fibres today.

Another characteristic related to the attenuation characteristic shown in the figure are the various windows of operation for optical communication systems. The first window, at

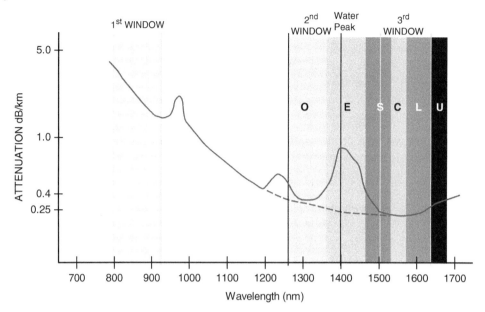

Figure 6.5 Wavelength-dependent attenuation.

800 nm, is associated with short-reach, multi-mode transmission where low-cost VSCEL lasers are available. The second and third windows, at 1300 and 1550 nm, are conventional windows of operation – 1550 nm (known as the 'C' band) targeting long-reach operations due to its low attenuation. An additional aspect of operation in the C-band window is the availability of optical amplifiers, based on erbium-doped fibre, that are capable of boosting the optical signal at specific points in the link.

6.4.2 Optical Fibre Dispersion

Another important mechanism that limits the overall capacity–reach performance of optical fibre systems is the fibre dispersion. The is an effect caused by the non-constant propagation time through a fibre for all spectral components of an optical signal. The three main causes of dispersion are:

- Material dispersion
- Modal dispersion – seen only in multi-mode fibres
- Waveguide dispersion – most significant in single-mode fibres and caused by the different velocities of the optical field between core and cladding materials

Figure 6.6 shows the contributions of waveguide and material dispersion to the total dispersion of single-mode fibre.

The impact of dispersion in an optical fibre system is to cause the various spectral components in the optical data stream to arrive at different times at the end receiver – having the effect of 'broadening' the signal and causing energy from one bit to fall into subsequent

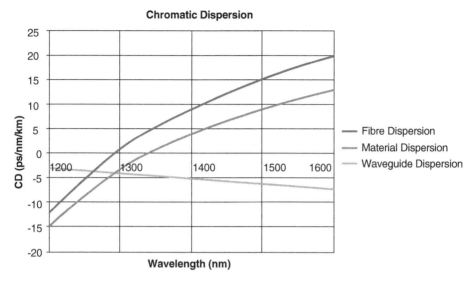

Figure 6.6 Dispersion characteristics of optical fibre [11].

bit periods. This causes distortion in the optical signal, and ultimately limits the reach that can be achieved. Higher-rate signals are more sensitive to this effect due to the reduced bit period and higher spectral components inherent in the signal.

6.5 TSN Transport Network for the Low-Layer Fronthaul

The IEEE 802.1 Standards Group has defined TSN in the IEEE 802.1CM and IEEE 802.1CMde standards [12, 13] to enhance switched Ethernet for time-critical applications for low-layer split cloud–RAN architectures, also known as fronthaul transport.

TSN is part of the IEEE 802.1 family of standards and is designed to provide deterministic forwarding on standard (Ethernet) switched networks. It is a Layer 2 technology that can be centrally managed (SDN) and use coordinated scheduling to ensure performance for real-time applications. Real-time deterministic communications are important to many industries (e.g. aerospace, automotive, transportation, utilities and manufacturing). In many of these industries, TSN is emerging as the baseline for real-time networking.

Published in 2018, the TSN fronthaul standard (IEEE 802.1CM) profiles IEEE 802 standards and ITU-T synchronization recommendations specifically for fronthaul transport. The standard specifies an Ethernet bridged network for low-layer split architecture connecting RU to a remote baseband (DU processes).

Synchronization work from the ITU-T's Study Group 15 was also referenced in developing IEEE 802.1CM. This study group has defined the network-based synchronization profiles for packet transport networks, including Ethernet in the fronthaul. The timing/sync section of the TSN profile describes the requirements for the time synchronization that is critical for TSN.

The forwarding of the TSN switching is based on the VLAN bridge specification in IEEE Std 802.1Q [14]. For fronthaul transport networks using TSN, the support of at least six different VLAN-IDs is required, including the default PVID (which is untagged by default, e.g. to carry non-fronthaul traffic). The Ethernet frames that are supported are RoE, eCPRI and other packets that are transported over the Ethernet infrastructure The TSN forwarder supports the following functionality in addition to those for VLAN-aware bridges:

- The use of at least six VIDs, one of which is the default PVID, configured to be untagged on all ports
- A minimum of three traffic classes on all ports, of which all traffic classes support the strict priority queuing algorithm for transmission selection
- Optional support for frame pre-emption

The TSN for fronthaul standard defines two transport profiles, both of which are applicable to CPRI and eCPRI protocols. Profile A forwards user data (IQ data) as a high-priority traffic class compared to the priority of the control and management data. Profile B includes components of Profile A but also adds a TSN feature called frame pre-emption (defined in IEEE 802.3br [15] and 802.1Qbu [16]) to prioritize different traffic types.

Frame pre-emption specification IEEE Std 802.1Qbu is the suspension of the transmission of a pre-emptable frame to allow one or more express frames to be transmitted before the transmission of the pre-emptable frame is resumed.

6.6 TDM-PONs

6.6.1 TDM-PONs as Switched Transport Network for Backhaul and Midhaul

A TDM-PON allows TN over a P2MP WS-ODN access network. It consists of a CO node, called an Optical Line Termination (OLT), and multiple user nodes, called Optical Network Units (ONUs) or also known as Optical Network Terminals (ONTs). Each TDM-PON takes advantage of WDM for BiDi working, using one wavelength for downstream traffic and another for upstream traffic on a single strand of single-mode fibre (ITU-T G.652).

TDM-PON is a shared bandwidth (in a single wavelength), as the OLT sends a single stream of downstream traffic that is seen by all ONUs. Each ONU only reads the content of those packets that are addressed to it. Encryption is used to prevent eavesdropping on downstream traffic. In the upstream direction a Time Division Multiple Access (TDMA) scheme is used in which ONUs are granted to send a traffic burst in a certain timeslot as scheduled by the OLT.

In an almost competing mode, the Institute of Electrical and Electronics Engineers (IEEE) and the Telecommunication Standardization Sector of the International Telecommunication Union (ITU-T) in the Full Service Access Network (FSAN) Working Group have developed standards for TDM-PON for FTTH. The IEEE Ethernet PON (EPON or 10GEPON) standard 802.3ah-2004 uses standard 802.3 Ethernet frames and is an implementation of the TN for BH/MH using 802.1 switches. On the other hand, the ITU-T G.984 Gigabit-capable Passive Optical Networks (GPONs) standard has a GPON Encapsulation Method (GEM) that allows very efficient packaging of 802.1 user traffic with frame segmentation. Therefore, the TDM-PON by ITU can also be seen as a TN for BH/MH with

802.1 network switching. ITU GPON networks have, since 2007, been deployed in numerous wireline access networks (FTTH) across the globe, and the trends indicate higher growth in GPON than other PON technologies. The ITU-T-specified TDM-PON also plays its part in the security challenge for TN, as this technology includes an advanced encryption standard with 128-bit keys (AES-128).

The need for next-generation PON (NG-PON) networks has emerged since 2020. For an operator with an existing ITU GPON network, the following next-generation PON technologies were defined.

- ITU-T G.987.3 [17] XG-PON (X = 10, G = Gigabit PON) was the first standardized next-generation technology. It delivers 10 Gbps downstream and 2.5 Gbps upstream (10/2.5G) using a single fixed wavelength in each direction.
- ITU-T G.989 [18] TWDM-PON was included as a NG-PON2 technology. It adds more wavelength pairs on the fibre (initially four channels or four pairs of wavelengths in the upstream and in the downstream, with four more channels possible in the future). TWDM-PON supports flexible bitrate configurations (2.5/2.5G, 10/2.5G and 10/10G) and uses tuneable lasers that allow operators to dynamically assign and change the wavelength pair of the four (later eight) in use. Hence, NG-PON2 is 40 Gbps aggregated as it has four channels of TWDM-PON.
- ITU-T G.9807.1 [19] XGS-PON (X = 10, G = Gigabit, S = symmetrical PON) delivers 10 Gbps in both directions but also supports dual-rate transmission. Dual rate means 10/10G XGS-PON ONUs and 10/2.5G XG-PON ONUs connected to the same OLT port through a native dual upstream rate TDMA scheme and a TDM scheme in the downstream.
- The Multi-PON Module (MPM) is also named COMBO, where GPON optics and XGS-PON optics are combined with a co-existence filter on the same small form factor pluggable. As such, an aggregated 12.5 Gbps capacity is available bidirectionally on a single fibre.
- An industry standard or MSA was set up to define a 25GS-PON in 2020 (25gspon-msa.org) with over 50 partners. Based on an IEEE wavelength plan, this TDM-PON (25G = Gigabit, S = symmetrical PON) delivers 25 Gbps in both directions but also supports dual-rate transmission. Dual rate means 25/25G 25GS-PON ONUs and 25/10/G 25G-PON ONUs connected to the same OLT port through a native dual upstream rate TDMA scheme and a TDM scheme in the downstream. As this MSA defines yet another wavelength pair (UW/DW) in the 1270–1370 nm O band, it is possible to co-exist GPON with XGS-PON and 25GS-PON on the same single fibre for a total aggregated capacity of 37.5 Gbps.
- The latest standards published are the ITU-T G.9804 series of higher-speed PONs, starting with 50G-PON which delivers $n \times 12.5$ Gbps line rate in both directions with different constellations of n. It uses either the upstream wavelengths of GPON or of XGS-PON, yet it is possible to co-exist XGS-PON with 25GS-PON and this 50G-PON on the same single fibre for a total aggregated capacity of 85 Gbps. The same common Transmission Convergence (TC) layer (or PON MAC) is specified for all. This higher-speed PON specification also plans for an improvement to optional AES-256 to ensure security in a quantum compute world.

All the above technologies are designed to be deployed co-existing on the same WS-ODN as GPON. These NG-PONs used other wavelength pairs and therefore can be wavelength multiplexed on the same ODN. It has been studied how a TDM-PON fits the switched TN for 5G in a shared brownfield infrastructure for backhaul and midhaul (higher-layer split gNB architectures with standard F1 interface). This analysis can be found in the ITU supplementary document G.sup.75 [17].

6.6.2 TDM-PONs as Switched Transport Network for Fronthaul

With the new series of TDM-PONs that have more than 10 Gbps capacity, the capacity challenge of a fronthaul connection is within reach. The 5G low-layer split fronthaul data rate should be traffic dependent and independent of the number of antennas, therefore supporting multiplexing via a switched 802.1 network.

The transport of a low-layer split RAN option (LLS-FH) has been profiled by IEEE TSN as described above. It is clear that the capacity of the TN is by far not the only technological challenge to transport LLS-FH with TDM-PON. Especially, the latency and time synchronization challenges will be important to address.

The ITU-T-defined TDM-PONs all use a time-distribution function with an accuracy based on the equalization delay of each endpoint (ONU) that a TDM-PON uses for ranging. This time-distribution function has the same accuracy as the IEEE 802.1AS [20] defined for IEEE EPON. Therefore, the higher-speed TDM-PON of ITU-T supports the accuracy required for a fronthaul cluster of base stations.

The ITU-T specifies, for higher-speed TDM-PONs, a number of low-latency features that ensure these higher-speed capacities can be used as transport for a low-layer split fronthaul as specified by O-RAN (Option 7.2x with Coordinated Transport Interface, CTI). The low-latency features include a coordinated dynamic bandwidth allocation for the upstream (CO-DBA). With classic DBA, a deterministic delay can only be achieved with constant allocation (all CIR). This CO-DBA should bring a deterministic delay in combination with a flexible (multiplexed) shared channel dynamic assignment.

6.7 Wavelength Division Multiplexing Connectivity

6.7.1 Passive WDM Architecture

These solutions expand fibre capacity, supporting multiple channels per fibre. In addition, the solutions are very reliable and power efficient because no power is needed for data transport. However, they lack transport protection, networking flexibility and robust OAM capabilities, which can lead to operational challenges. They also lack the clear network demarcation points needed between an MNO and the TN provider, thus limiting their applicability. For cell sites with a low number of O-RUs, the simplicity and cost of a passive solution provides significant benefits.

The passive WDM solution is based on the E2E all-passive method without optical amplification or dispersion compensation. The WDM technologies are defined based on the optical spectrum characteristics. The optical fibre spectrum has been further subdivided into operating bands O, E, S, C, L and U, as shown in Figure 6.5. Within these bands, groups of wavelengths with fixed wavelength spacings have been selected, with the designation CWDM (Coarse WDM), LWDM (LAN WDM), MWDM (Medium WDM) and DWDM (Dense WDM). Each group uses specific transmission technologies, optimized for number of channels, transmission bit rate, transceiver power, distance and cost points.

i) CWDM operation. 18 wavelengths are available over bands O to L, with 20 nm spacing between wavelengths (18 wavelengths are available with water peak 'flattened' fibre). The wide wavelength window for each channel allows uncooled, directly modulated transmission sources to be used with the large temperature fluctuations found in outside plant situations – making a CWDM solution very cost effective for access and C-RAN fronthaul deployments, especially for lower bit rates (up to approximately 10 Gbps). For 25 Gbps and higher, fibre dispersion limits the reach, and the number of channels that can be effectively used is reduced to 6.

ii) LWDM and MWDM operation. These are two WDM grids recently introduced to target passive 5G fronthaul and small cell applications requiring outside plant operation with capacities of 25 Gbps or higher and utilizing the low-dispersion window of the O band. LWDM supports 12 wavelengths between 1269 and 1332 nm, while MWDM was designed to operate over the existing 20 nm-spaced CWDM grid by creating two channels per CWDM window, spaced ±3.5 nm from the window centre. The key benefit of this technology is the high tolerance to dispersion with 25 Gbps channels, using low-cost transmitter technologies.

iii) DWDM operation. The minimum loss windows of the C and L bands are used for DWDM operation. The availability of optical amplifiers in these windows makes these suitable for longer-reach applications such as the midhaul and backhaul/core. The wavelength grid for operation in the C and L bands has been standardized by the ITU – providing three levels of capability.

These wavelength-based systems offer a good combination of characteristics for fronthaul transport.

The use of WDM entails deploying WDM optics at cell sites and using optical filters to combine multiple client (e.g. CPRI/OBSAI/Ethernet) signals (different, standardized wavelengths) onto a single optical fibre pair. These individual wavelengths are broken out again by similar WDM filters at the central location (see Figure 6.7).

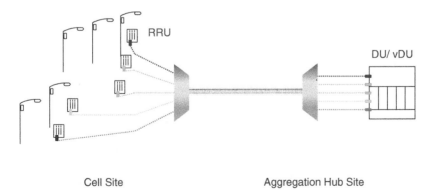

Cell Site   Aggregation Hub Site

Figure 6.7   Passive WDM.

The RU uses the fixed or tunable optical transceivers directly, connected to the passive multiplexer/demultiplexer through the branch optical cable. At the DU side of the CO, the passive multiplexer/demultiplexer performs wavelength multiplexing/demultiplexing, which realizes the one-to-one optical wavelength connection. Fully passive WDM solutions enable highly reliable, cost-effective CPRI/OBSAI/eCPRI transport, while meeting the strict latency and jitter requirements of these protocols. Passive WDM solutions having no active electronics and relatively simple optics at both ends can be deployed in all outdoor environments, providing significant CAPEX/OPEX savings. There is no requirement for battery back-up, and no power consumption. WDM technology provides a cost-effective transport option necessary to support the projected RRH and small cell rollouts.

The equipment at the O-RU and O-DU side may contain tunable optical transceivers with an integrated, in-band communication channel using technology such as subcarrier modulation. Known as smart, self-tunable transceivers, they allow the remote transceiver, embedded in the O-RU, to self-tune to the transceiver wavelength connected at the central site without the use of an external supervisory channel, or the need to administratively enable the host equipment.

The latency challenge for fronthaul described earlier is also well supported by using passive WDM. The lack of additional signal framing along the optical path between the RRH/RU and the BBU/CU means that the only source of transport-incurred latency is due to signal propagation through the passive filters and fibre. This allows the MNO to maximize the distance between the radios and the centralized hub site. Because there is no alteration of the CPRI/OBSAI/eCPRI signal, fronthaul solutions are transparent to the wireless system and will not affect the radio performance if the maximum fibre lengths are respected.

6.7.2 Active–Active WDM Architecture

Active–active WDM solutions deploy an active element at each end of the fronthaul link with the ability to multiplex or aggregate multiple RU ports onto one line fibre. They also include OAM capabilities, enabling network visibility and management of the TN in order to provide demarcation and SLA performance management. And they can simplify and accelerate deployments by using translators that convert black and white wavelengths to WDM wavelengths, thereby avoiding impacts to the RAN equipment or installation procedures. The key disadvantage with the active solution is the need to provide a powered remote site to house the active equipment – entailing long cycle times for planning, regulatory approval and certification and additional space lease, power, management and maintenance expenses over the lifetime of the deployment. For cell sites with a higher number of RUs, the aggregation and multiplexing capability of the active solution provides significant benefits.

The active WDM solution uses outdoor active WDM equipment for electrical and/or optical-layer multiplexing at both the remote station and aggregation hub (central) office, as shown in Figure 6.8. Active packet solutions provide an adaptation to an Ethernet packet protocol that can then be aggregated by a switched network (as defined by the TSN fronthaul profile described above) onto a higher-rate network interface (100 Gbps or more). The main advantage is the ability to multiplex the traffic streams from the cell site onto low-cost 100G transceivers – obtaining very low transport cost-per-bit fronthaul connectivity.

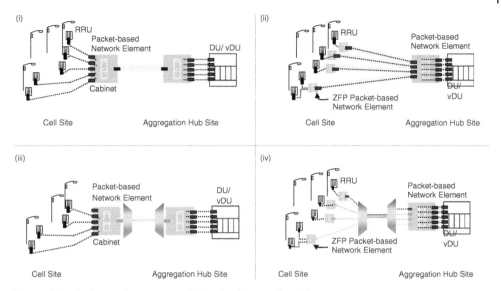

Figure 6.8 Active–active packetized fronthaul network architectures.

Active WDM solutions can also support longer transmission reach (10–30 km), higher service availability with full duplex transport protection and more flexible networking (e.g. point-to-point, linear aggregation and ring architectures).

Two active–active WDMs solution are feasible: (a) an active transparent architecture which provides recolouring of the interfaces from the RRUs to the WDM interfaces, which are then optically multiplexed onto a single fibre (pair) towards the central site; and (b) an active packet architecture, allowing for Common Public Radio Interface (CPRI) traffic Ethernet packetization. Packetization through an IEEE 1914.3 radio over Ethernet structure-aware mapping of the TDM-based CPRI signals can further reduce the required transport capacity, as it can remove unused antenna carriers (AxC) in the CPRI stream, and furthermore the packets from other applications (including the eCPRI stream) can be statistically multiplexed. When the total transport capacity is below 100 Gbps, typically grey optical transceivers can be used for the majority of sites, and 100 Gbps DWDM transceivers are deployed only when the cell-site capacity exceeds this.

Typically, active packet-based equipment at the cell site can be deployed based on two different form factors – a cabinet-mounted 1RU pizza box-style unit or zero-footprint (ZFP) pole-mounted outdoor style unit. Figure 6.8 illustrates the different active–active packet-based architectures including (ii) and (iv), showing 100 Gbps packetized transport in a ZFP form factor at cell sites, and (i) and (iii), showing 100Gbps packetized transport in a cabinet form factor.

Active WDM chains can be formed by daisy-chaining a number of ZFP cell sites before the aggregated traffic is sent to the aggregation hub site. There are two solutions considered: (I) all connectivity is achieved via 100 Gbps grey optical interconnection and (II) the aggregated traffic is using 100 Gbps DWDM towards to hub while the intracell site connections are via grey optics, as illustrated in Figure 6.9.

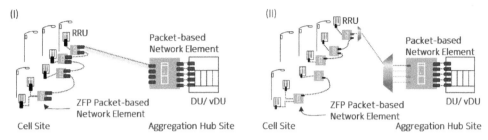

Figure 6.9 Active optical packetized WDM ring fronthaul network architectures.

6.7.3 Semi-Active WDM Architecture

The semi-active WDM solution is a simplification of the active WDM solution and an enhancement of the passive solution. The solution maintains a passive remote site with fixed or tunable pluggable modules installed in the RUs and multiplexed onto a single fibre towards the central site with a passive optical filter. At the central DU site, an active unit is placed to provide aggregation, demarcation, monitoring or fault detection and isolation functions.

The semi-active WDM solution not only greatly relieves the pressure of optical fibre resources, but also has advantages in terms of cost (compared with the active solution), management and protection outside the optical transceiver and O-RU/O-DU host systems (compared with the passive solution). It helps operators to build 5G fronthaul networks with low cost, high bandwidth and fast deployment.

The proposed semi-active WDM schemes are illustrated in Figure 6.10.

The semi-active WDM type I equipment should support query, configuration and sending OAM information. To further reduce system costs, the semi-active WDM type II equipment can perform simplified management, including sending the OAM information of optical modules to the active WDM equipment. With the accelerated deployment of 5G networks, the fronthaul network will have thousands of nodes, creating a need for maintainable management capability of the network.

The active WDM equipment can send management requests to the O-RU and manage the WDM optical pluggable modules in the O-RU, including query and configuration. The WDM optical modules in the O-RU can receive management requests from the active WDM equipment and then send the OAM information of optical modules to the active WDM equipment, including the wavelength and output power of the transmitter. The WDM optical modules in the O-RU and O-DU can send the OAM information of optical modules to the active WDM equipment automatically or at regular times as the optical modules are powered on. The WDM optical modules can add the OAM information onto the service signals and transport together in the same optical channel. The detection unit in the active WDM equipment can demodulate the OAM information, obtain the transmission performance of O-RU and O-DU, and then report it to the control system through the standard southbound interface.

The WDM-PON as defined by ITU-T in G.9802.1 is a semi-active WDM solution with WR-ODN as full passive outside plant elements and pluggable optical modules with tunable optics. Each channel provides bidirectional connectivity through a pair of wavelengths constituting a channel pair. Each WDM channel has a symmetric nominal line rate

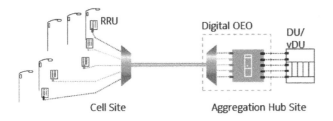

Semi-active WDM type I

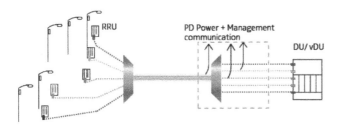

Semi-active WDM type II

Figure 6.10 Semi-active WDM.

combination of 25 and 10 Gbps per wavelength channel. The OAM channel is specified here as an Auxiliary Management Control Channel (AMCC) that, amongst other OAM features, controls the tuning of the optics. As a TN for low-layer split architecture (fronthaul) is deterministic and latency sensitive, this limits the maximum fibre distance classes up to 10 km. This definition is a semi-active WDM of Type 2.

The TDM-PON [6–6 above] over shared brownfield WS-ODNs uses the semi-passive architecture where the ONU is implemented as a pluggable stick supporting only one TDM-PON channel, saving space and minimizing power supply requirements. With 25GS-PON defined by an MSA or the G.9803 [16]-defined higher-speed PON, the 25G symmetrical channel is shared by all RUs on the same WS-ODN. This would be classified as a semi-active TN of Type 1 with only one channel pair in the original wavelength band (O band).

A combination of TDM and WDM (TWDM-PON) as defined for four channel pairs of 10 Gbps in the ITU-T NG-PON2 (G.989) is not yet specified for multiple channel pairs of 25 Gbps (or higher). Experience has shown that for a semi-active version, a pluggable remote optic is needed in the RU that requires considerations on optical penalty, tunable optics over four channel pairs in DWDM and a management channel.

6.8 Total Cost of Ownership for Fronthaul Transport Networking

Various industry studies, for example see Ref. [21], comparing the several fronthaul architectures from a TCO perspective have shown that for the low number of small cell clusters and low small cell interfaces needing to be aggregated to a centralized hub site, the passive-only solutions are the lowest cost for CWDM, MWDM and DWDM. This is directly related to the wavelength count in the fibre. This result is not surprising given that the three WDM technologies have been deliberately developed by the industry, to target this cost vs capacity trade-off.

Once the number of ports per small cell site or the number of cell sites that need to be aggregated becomes more significant, the active–active packet solutions have the lowest cost of ownership. Furthermore, for the 25 Gbps client interfaces and low cell sites per cluster (<10 approximately), ZFP architectures are the most cost effective. This is due to the relatively high cost of the powered cabinet required to house the remote active unit, amortized over the low number of cells. Within this region, for a low number of ports per cell, it is observed that ZFP chained architectures are the most cost effective, due to the benefit of aggregation of the per-cell traffic onto fewer network interfaces. When the capacity per cell is too large, the benefit of chaining diminishes and a ZFP DWDM architecture is superior. For a larger number of cell sites per cluster, the value of an active-packet cabinet-based solution can be seen, in which the statistical multiplexing from the large number of cells reduces the transport bandwidth and overall cost points. Figure 6.11 attempts to provide a general illustration of the discussion above, showing the regions of lowest TCO as a function of cell-site ports and cluster density.

Equally for the TDM-PON technologies, industry studies have shown the shared and dedicated leasing cost structure of the FTTH ODN and the impact on the TCO of two LLS-FH 3GPP options (Option 7.2x and Option 7.3). These studies show the impact of the shared vs dedicated models in relation to the TDM PON transmission rate, concluding that there is an up to 50% improved cost-effectiveness benefit of a TDM-PON technology in comparison to a WDM-PON or semi-active DWDM solution. Sensitivity analysis has also shown an appropriate TCO zone of applicability for this technology, relative and very sensitive to the mobile radio parameters of carrier BW and MIMO (Multiple Input Multiple Output) layers.

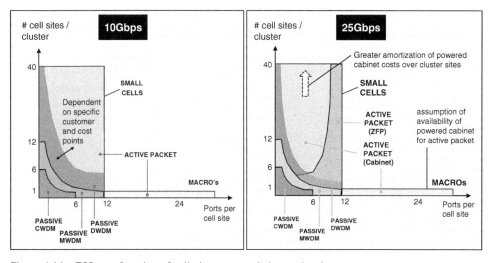

Figure 6.11 TCO as a function of cell-site ports and cluster density.

The benefits of the active solutions are evident, in particular where the statistical multiplexing gain from a large number of cells can significantly reduce the transport bandwidth. However, the burden of a powered location for the active equipment can be significant and leads to a passive or semi-active solution being more cost effective, especially when the number of cells per cluster and/or the fibre lease costs are low. The results of the TCO studies were derived for a given set of assumptions, but it is noted that in practice, these can change substantially from operator to operator and should be evaluated based on each operator's specific conditions.

References

1 NGMN Overview on 5G RAN functional decomposition, IOWN – requirements of supported data transport.
2 3GPP TS 22.261 V18.3.0 Service requirements for the 5G system (2021-06).
3 ITU-T G.8031 Ethernet linear protection switching.
4 ITU-T G.8032 Ethernet ring protection switching.
5 IEEE 802.1ag 802.1ag – Connectivity fault management.
6 ITU-T Y.1731 OAM functions and mechanisms for Ethernet-based networks.
7 MEF 35.1, MEF 35.1 Service OAM performance monitoring implementation agreement.
8 MEF 30.1 Service OAM fault management implementation agreement: phase 2.
9 MEF 17 Service OAM framework and requirements.
10 IEEE P1914.3 Standard for radio over Ethernet encapsulations and mappings.
11 Fibre characterization testing for long haul, high speed fibre optic networks. https://www.thefoa.org/tech/ref/testing/test/CD_PMD.html.
12 IEEE 802.1CM Time-sensitive networking for fronthaul.
13 IEEE 802.1CMde Time-sensitive networking for fronthaul, amendment 1: enhancements to fronthaul profiles to support new fronthaul interface, synchronization, and syntonization standards.
14 IEEE standard for local and metropolitan area networks – bridges and bridged networks (IEEE Std 802.1Q).
15 IEEE standard for Ethernet amendment 5: specification and management parameters for interspersing express traffic (IEEE Std 802.3br-2016).
16 IEEE standard for local and metropolitan area networks – bridges and bridged networks – amendment 26: frame pre-emption (IEEE Std 802.1Qbu-2016).
17 ITU-T G.987.3: 10-Gigabit-capable passive optical networks (XG-PON): transmission convergence (TC) layer specification.
18 ITU-T G.989: 40-Gigabit-capable passive optical networks (NG-PON2): definitions, abbreviations and acronyms.
19 G.9807.1: 10-Gigabit-capable symmetric passive optical network (XGS-PON); recommendation.
20 IEEE 802.1AS timing and synchronization for time-sensitive applications.
21 Levrau, L. and Remedios, D. Guidelines for a cost optimised 5G WDM-based fronthaul network. 2021 European Conference on Optical Communication (ECOC).

7

Wireless Backhaul and Fronthaul

Paolo Di Prisco, Antti Pietiläinen and Juha Salmelin

Wireless transport represents a cost-optimized solution complementing fibre for mobile backhaul, especially for tail and first aggregation level, as well as for enterprise, mission critical and vertical markets. Microwave links are a strategic asset for 5G deployment, while Free Space Optics (FSOs) are an option only for very specific use cases.

The ongoing technological evolution to support different mobile network scenarios and requirements is promising, as 5G microwave is projected to reach up to 100 Gbps in future transport capability.

7.1   Baseline

As of today, 65% of cell sites are backhauled through microwave links out of the Far East region (Korea/Japan/China), where fibre is widely available. In other regions (mainly Europe, the Middle East, Africa and India as examples), wireless transport is providing the means for current mobile base stations (2G/3G/4G and early 5G) to be connected to the core network. Moreover, in urban areas, if the Communications Service Provider (CSP) is not incumbent, fibre leasing may be too expensive, especially in view of the evolution towards higher and higher speed.

For these reasons, wireless systems are expected to be one of the major transport solutions in the future, as reported by different analysts (see Figure 7.1). There is even additional expected business in the coming years, generated by new use cases.

Line-of-sight microwave systems balance three factors (throughput, reach and availability) leveraging on different frequency bands, each with different performance characteristics.

In general, the higher the frequency, the larger the available channel spacings and therefore the higher the capacity and the lower the transport latency. Traditional systems use lower-frequency microwave (6–42 GHz) spectrum bands, while in recent years E-band (80 GHz) is becoming a relevant option to provide multi-Gbps throughput in small form factor (see Figure 7.2).

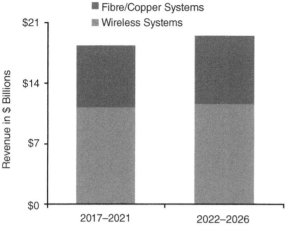

Figure 7.1 Macro-cell backhaul transport (Microwave Transmission and Mobile Backhaul 5-Year Forecast Report, Dell'Oro, January 2022).

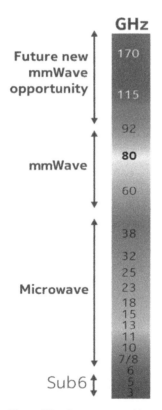

Figure 7.2 Spectrum used in wireless transport.

As described in the following sections, it is also possible to combine E-band with traditional bands to achieve longer distances and preserve high availability for the most valuable traffic.

With various regional peculiarities, microwave is used to serve different connectivity scenarios. In order to have an idea about the distance which is covered in today's deployments, some Nokia data about cumulative link length distributions worldwide are reported below, even if there is a general trend towards shorter links moving forward (also due to fibre penetration):

- 40% below 4 km
- 30% between 4 and 8 km
- 20% between 8 and 16 km
- 10% above 16 km

This needs to be taken into account as a basis for the 5G evolution upgrade, as many access sites still do not have a fibre point of presence close by.

7.2 Outlook

Looking at the future, the microwave industry is going to expand to even higher frequencies with W-band (92–114.25 GHz) and D-band (130–174.8 GHz) solutions.

While the W band is today viewed as a likely extension or spare of the E-band because of its similar propagation behaviour, D-band technology can be really disruptive, promising ultra-high capacity (up to 100 Gbps) at ultra-low latency (<10 μs) and small form factor, with antennas down to a few square centimetres in size. That is primarily an option for short-reach transport in urban environments.

Technology is evolving also for systems up to 86 GHz and for associated networking capabilities. This will be better described in the following paragraphs. At a high level, the combination of traditional and new frequency bands can address 5G's challenging requirements, as well as enterprise and mission-critical applications (e.g. public safety use cases) which require high-resiliency network performance.

With respect to this, it should be noted that fibre is also subject to possible failures and redundancy, with wireless connectivity not an unusual scenario.

Current and future systems will include:

- Traditional microwave bands which can be aggregated to provide multiple Gbps
- mmWave bands below 100 GHz, providing 20 Gbps with one-way latency around 10 μs
- Higher bands with the promise of reaching 100 Gbps and smaller form factor (with active antennas as well)

As shown in Figure 7.3, the spectrum is the main fundamental asset on which operators can leverage in order to plan for the wireless transport network evolution, therefore it is fundamental that new bands will be assigned in the future by administrations and that the industry will develop techniques which enhance their efficiency.

Market analysts are forecasting for the next years a slight growth in the microwave industry (and backhaul in general) driven by 5G and small cells (see Figure 7.4).

The microwave transmission market is estimated to reach around $3.4bn by 2024, of which a relevant part will be represented by 5G networks. The value of enterprise/vertical applications is expected to be non-negligible, at around 1B$.

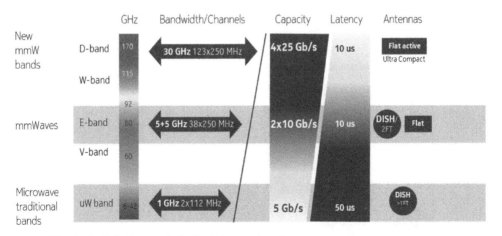

Figure 7.3 Radio link characteristics by frequency band.

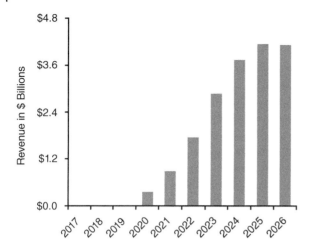

Figure 7.4 5G mobile BH transport revenue (Microwave Transmission and Mobile Backhaul 5-Year Forecast Report, Dell'Oro, January 2022).

While initial and sparse 5G deployments mostly occur at cell sites which are already served by fibre, it is then forecast that wireless backhaul will be used in upcoming rollouts by incumbent operators. Moreover, there is also a tendency for Tier 2/3 operators (not owning a converged network) to use more microwave in future.

On top of traditional backhaul, new transport interfaces are emerging (as described previously); in particular, midhaul connectivity (between CU and DU) can also be implemented through microwave, while fronthaul will represent a new potential business exploiting high-frequency bands (see Figure 7.5).

The backhaul spectrum cost is usually much lower than the access spectrum one, but in the future more and more synergies will be allowed, especially when an operator owns spectrum blocks, thus being able to drive the implementation on different technologies.

Spectrum fees play a significant role in determining the economical applicability of wireless solutions, as they influence the operator business case and consequent choices. As an example,

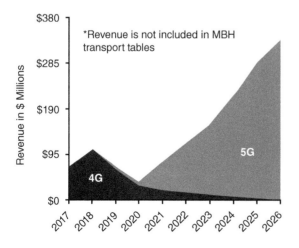

Figure 7.5 Mobile fronthaul (CPRI/eCPRI) transport (Microwave Transmission and Mobile Backhaul 5-Year Forecast Report, Dell'Oro, January 2022).

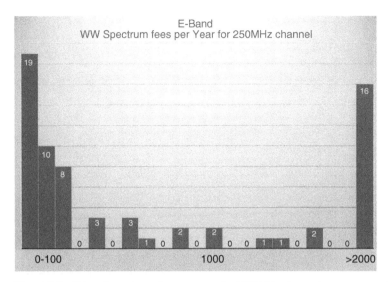

Figure 7.6   E-band spectrum fees per year for 250 MHz channel.

E-band usage is favoured in most countries by a light licensing scheme and a relatively low annual fee. This is allowing a fast-growing 5G backhaul network where fibre is not present.

In Figure 7.6 we report the E-band spectrum fee level (per 250 MHz channel, per year) in different countries. Most of the cases are below 300€, which is a quite limited level, thus boosting technology adoption. Different rules are then applied country by country in case of cross-polar channel usage or larger spectra, but in general the average cost per Gbps favours E-band over traditional frequencies.

The success of future spectrum bands used for wireless backhaul will also depend on the licensing scheme and associated costs, which will be determined by the different administrations. Organizations like ETSI ISG mWT (millimetre Wave Transmission) are devoted to analysing and promoting the usage of various wireless technologies for mobile transport.

7.3   Use Cases Densification and Network Upgrade

Different geographical use cases may be applicable, depending on the baseline network and the RAN upgrade type that is undertaken.

In ultra-dense urban areas or hotspots such as crowded squares, airports and stadiums, 5G networks will be deployed with the radio access millimetre wave layer (26/28/39 GHz). Very high-capacity backhaul is needed (10 Gbps and above) and the transport link lengths are less than 1 km in most cases. The E-band and frequencies beyond 100 GHz are addressing this need.

In the urban/suburban scenario (up to 7–10 km link distance), the access layer will be based mainly on sub-6 GHz frequencies with connectivity requirements that are still quite demanding in terms of capacity (5–10 Gbps). In this case, Carrier Aggregation (CA) between microwave and mmW carriers will be the main solution on the backhaul side.

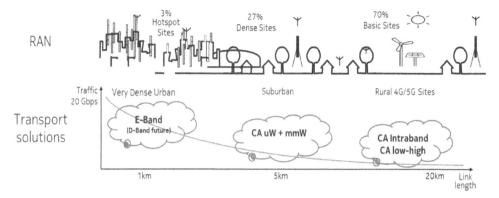

Figure 7.7 Transport solutions vs geographical area.

In rural settings, where the geographical area coverage is larger, the access network will be mostly based on frequencies below 1 GHz. The transport network will need to backhaul up to a few Gbps and the link lengths commonly exceed 7–10 km. Low-frequency band combinations (CA) are the way in which these requirements can be satisfied (see Figure 7.7).

In terms of transport network evolution, two main scenarios can be envisioned.

1) Existing network upgrade. This first scenario is mainly related to macro-cells backhaul, which represents the majority of the market. It leverages on existing 2G/3G/4G sites which need to be upgraded to 5G (NSA architecture in the first step). When microwave is already used as backhaul technology (especially in rural areas or for Tier 2/3 CSPs), such microwave systems need to be upgraded to higher capacity to sustain the new requirements.

Bringing high capacity to rural sites is one of the 5G challenges, as plenty of installations reside in places that are quite difficult to reach. Nevertheless, specific microwave solutions for long-haul applications have also evolved to provide high throughput for long distances, provided that a spectrum in low microwave bands (6–11 GHz) is available.

2) Densification. New opportunities arise from the expected 5G densification, which calls for small cells (including 5G mmW). Following the densification network trends previously described, the main involved use cases are therefore:

   • Urban densification with street to rooftop or street to street connectivity, exploiting poles and street furniture
   • High-throughput aggregation rooftop to rooftop

This is the typical scenario for which a fibre point of presence might be a few hundred metres away from the radio access point. In such a case, TCO (Total Cost of Ownership) evaluation tends to favour microwave connectivity as fibre trenching costs are unlikely to reduce over time. Moreover, in dense urban environments it is common to have fibre access at building level, but not at street-pole level.

All x-haul types (backhaul, midhaul, fronthaul) are concerned, depending on the deployment scenario, including small cells, 5G mmWave and RRH (Remote Radio Head) connectivity. In particular, the increase of cloud-based RAN architecture, with baseband pooling capabilities, enables Ethernet-based fronthaul (eCPRI) wireless transport.

Key requirements for such a scenario are:

- Reduced form factor for optimal visual impact and reduced pole weight suitable for urban furniture (Figure 7.8)
- High throughput, especially for eCPRI transport
- More connections but shorter on average

In order to fulfill these needs, a high-frequency spectrum is the most suitable solution: E-band and, in the near future, a spectrum above 100 GHz (e.g. D-band) can satisfy very high-capacity requirements in small form factor (see Figure 7.9).

As a summary, wireless solutions and associated technology advancements can satisfy both scenarios (network upgrade and densification).

Figure 7.8 Wireless backhaul installation example in urban area.

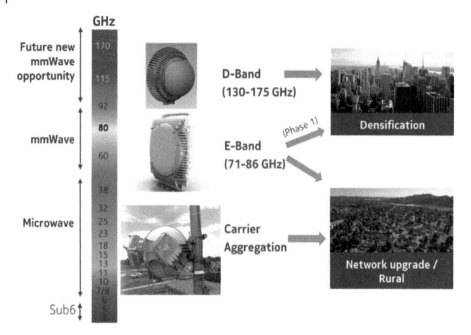

Figure 7.9 Use cases vs frequency bands.

7.4 Architecture Evolution – Fronthaul/Midhaul/Backhaul

RAN decomposition and densification will require the transformation of transport connectivity from simple hierarchical tree topologies (classical backhaul, connecting the access points to the core) to a more complex ring/meshed infrastructure. This new concept (fronthaul/midhaul/backhaul convergence) will serve a variety of use cases within the same network, as shown in Figure 7.10.

Such transport network evolution and the demanding 5G targets will dramatically increase capacity, connectivity and agility requirements.

Microwave systems are going to address the following architecture options:

- Backhaul/midhaul, mainly for macro cell (connecting DU and CU) and for 5G mmW used as densification
- Fronthaul, mainly for small cells and in general enabled by eCPRI, which provides significant capacity benefits with respect to legacy fronthaul solutions

7.5 Market Trends and Drivers

In this paragraph, the main market trends for the next years are described, driving the technology evolution for microwave systems. Clearly, the main Key Performance Indicators (KPIs) (capacity, latency, reliability) represent the main drivers to upgrade a transport network for an operator.

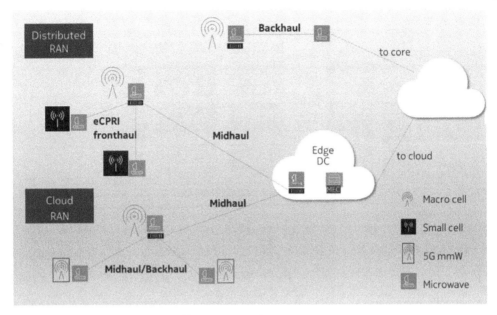

Figure 7.10 Wireless transport architecture scenarios.

As far as microwave is concerned, the main trends are as described in the following.

7.5.1 Data Capacity Increase

The most straightforward trend is the capacity increase due to end-user requirements. It is already ongoing with the shift of all cell sites from 1G to 10G technology, which implies line interface evolution (towards 25G) and more powerful switching capabilities usually provided by microwave networking units or cell-site gateways/routers.

From the transport radio perspective, the requirements have moved rapidly towards 10 Gbps backhaul aggregation capabilities on a single link, and in some cases up to 20 Gbps, especially when chaining of multiple cell sites is concerned. With the advent of Ethernet fronthaul, future wireless systems will reach up to 100 Gbps, especially in dense urban environments.

Figure 7.11 shows a summary of the main transport requirements and corresponding solutions related to 5G backhaul/midhaul, where it can be seen how wireless transport is a good fit in the different use cases.

In a similar way to what is occurring on the access side, this capacity growth is associated with a general shift towards increased usage of higher-spectrum bands. mmW bands are currently growing at around 25% CAGR, as depicted in Figure 7.12.

Such a move to a higher spectrum is coherent with the expected network evolution, which foresees shorter (on average) microwave links to cover the connectivity towards the fibre point of presence.

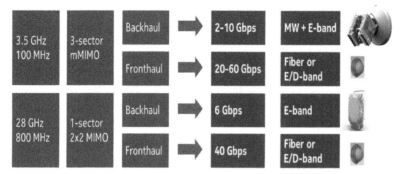

Figure 7.11 5G requirements and wireless transport capabilities.

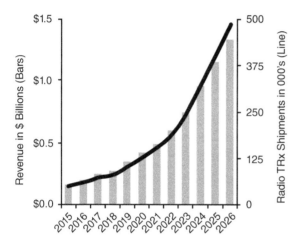

Figure 7.12 E/V band radio links growth forecast (Microwave Transmission and Mobile Backhaul 5-Year Forecast Report, Dell'Oro, January 2022).

7.5.2 Full Outdoor

With the advent of 5G, network densification will diminish the need for large hubs and will require more connectivity in dense urban environments. These factors are pushing the CSPs and enterprises to prefer more and more full-outdoor (FODU) solutions, which do not require an indoor box (and shelf) to operate but can be directly connected to router or base station (Figure 7.13).

This trend is confirmed by analyst reports, showing an acceleration of FODU deployment, which will quickly recover the gap vs the popular split-mount configuration.

7.5.3 New Services and Slicing

In the future, the transport network will have to support diversified requirements depending on the service type. Moreover, 5G advancements will allow us to satisfy challenging KPIs, such as very low latency and ultra-high reliability for the URLLC use case (Figure 7.14).

Transport slicing (over multi-technology transport, including microwave) can be put in place in order to logically (or physically) isolate different services and provide the needed requirements. To satisfy that, microwave equipment and network management systems

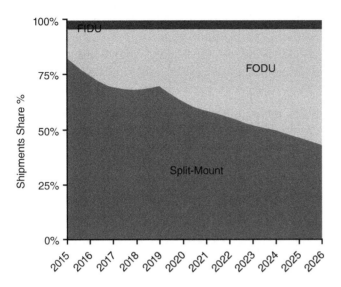

Figure 7.13 Microwave transmission radio TRx shipment share (Microwave Transmission and Mobile Backhaul 5-Year Forecast Report, Dell'Oro, January 2022).

URLLC scenario	E2E latency	Reliability	User data rate
Discrete automation (motion control)	1 ms	99.9999%	1 to 10 Mbps
Process automation (remote control)	50 ms	99.9999%	1 to 10 Mbps
Process automation (monitoring	50 ms	99.9%	1 Mbps
Electricity distribution	5 ms	99.9999%	10 Mbps
Intelligent transport systems	10 ms	99.9999%	10 Mbps
Remote control	5 ms	99.999%	<10 Mbps

Figure 7.14 URLLC requirements (3GPP).

are continuously evolving, with Quality of Service (QoS) enhancements for SLA enforcement, enhanced troubleshooting and network assurance.

7.5.4 End-to-End Automation

Virtual network functions, edge cloud and dynamic network slicing are only some of the changes future networks will have to manage. With this upcoming network complexity, it is becoming increasingly important to exploit automation tools in every possible domain (access, transport, core), thus including the microwave backhaul section.

Most operators are willing to include end-to-end SDN capabilities in the network, in order to reduce operation costs, scale resources and automate network service provisioning and assurance. Transport will have to become truly flexible and programmable. For this reason, different features in such domains are available in today's microwave network equipment and are expected to evolve to cover additional use cases. This includes:

- IP/L3 networking, to streamline e2e transport networking and increase reliability
- Carrier SDN service automation over the whole transport chain, using real-time telemetry for advanced traffic engineering and automatic provisioning
- Other specific SDN use cases to increase energy and spectral efficiency
- Advanced analytics
- Combined RAN and transport AI/ML use cases (e.g. traffic rerouting based on transport load/energy efficiency)

7.6 Tools for Capacity Boost

7.6.1 mmW Technology (Below 100 GHz)

In recent years, microwave transport has been expanding towards higher-frequency bands, thanks to technology evolution enabling new capabilities. In particular, E-band (71–76/81–86 GHz) technology has evolved from initial 1 Gbps-capable devices, launched more than a decade ago, to new and more powerful equipment, capable of 10/20 Gbps in the range of a few kilometres (2 GHz spectrum). Crosspolar operation (using both H and V polarizations) allows us to significantly extend coverage or double capacity at the same link distance.

Low visual impact is a key factor to facilitate deployment in dense urban environments. This is also leading to antenna integration within the radio unit, achieving a lower-TCO solution suitable for pole installations.

E-band addresses multiple cases due to its versatility, as in Figure 7.15, from urban to dense urban environments (with small form factor option), to suburban for a longer link length in aggregation with traditional microwave band, as described in the next paragraph.

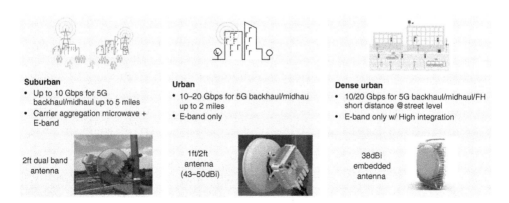

Suburban
- Up to 10 Gbps for 5G backhaul/midhaul up to 5 miles
- Carrier aggregation microwave + E-band

2ft dual band antenna

Urban
- 10–20 Gbps for 5G backhaul/midhau up to 2 miles
- E-band only

1ft/2ft antenna (43–50dBi)

Dense urban
- 10/20 Gbps for 5G backhaul/midhaul/FH short distance @ street level
- E-band only w/ High integration

38dBi embedded antenna

Figure 7.15 Use cases for E-band deployment.

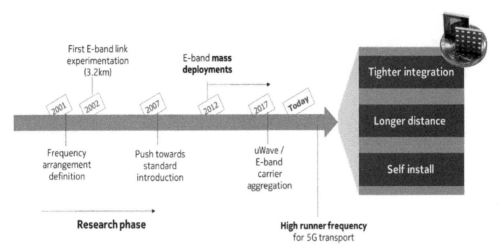

Figure 7.16 E-band success story.

Such a spectrum represents a success story for wireless transport. After the initial frequency arrangement definition (20 years ago), the first experiments have already addressed a remarkable distance, under a research wave already envisaging positive future prospects for such a band.

Following proper standardization activity, E-band has been successfully proposed as a backhaul solution and today represents about 15% of the total addressable market, being used also in early 5G deployments.

The future of such a spectrum is quite bright, as this is going to become the most utilized spectrum for wireless transport, thanks to its capability to deliver high capacity and very low latency. Technology is evolving to satisfy more powerful needs (Figure 7.16), such as:

- Even smaller form factor for full densification, thanks to tighter integration among RF components (single chip), as well as between RF and antenna
- Longer distance, with increased transmit power and stabilized antenna systems, to capture new use cases or increased availability
- Self-installation, to reduce network operation costs, thus avoiding manual work to set up the wireless point-to-point link

7.6.2 Carrier Aggregation

One of the most powerful tools to increase capacity/availability performance is CA, which combines different channels and spectrum bands over the same link. This is also a solution allowing us to reduce TCO and tower load, because the trend – which has already started – is to provide such a combination in a single box and with a single antenna supporting multiple-frequency bands (Figure 7.17).

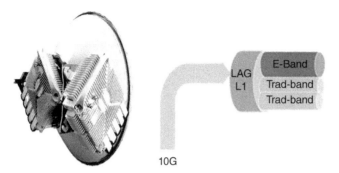

Figure 7.17 Carrier aggregation in radio links.

Depending on the combination type, different 5G transport scenarios can be covered:

- In an urban scenario (or dense areas), CA can be used to boost capacity to multiple tens of Gbps.
- In a suburban environment, the combination of microwave and mmWave carriers allows an optimal balance for a reliable and high-capacity transmission.
- In lower-density areas, uWave bands with different characteristics can be combined to increase capacity and distance.

Currently the most popular use case is the aggregation of microwave and mmWave: by combining the E-band channel (e.g. up to 2 GHz) with a traditional microwave frequency band channel (generally up to 56–112 MHz between 15 and 23 GHz), it is possible to achieve longer distances (up to 10 km) and preserve the usual high availability for the most valuable traffic. The E-band spectrum would thus provide the capacity boost needed by 5G networks (Figure 7.18).

This is a scenario many operators are evaluating in order to upgrade the current microwave network, as the existing deployments can easily be put in overlay with the new E-band equipment in order to meet the above objectives.

An emerging use case is the one combining two different microwave frequency bands, allowing us to extend the link distance and capacity. The main concept is that the lower-frequency bands permit us to cover a longer distance (thanks to better propagation) with the required availability for high-priority traffic, while the higher band adds additional capacity.

Taking as example a European region (associated with an average rain rate, defining its average propagation statistics), while using only a high microwave band (e.g. 38 GHz) in a link will enable us to reach a few kilometres, combining it with a lower-frequency band (e.g. 13 GHz) will allow us to cover a distance up to almost 20 km.

Figure 7.19 shows an example of how much we can extend the link distance (to cover new scenarios), combining the frequency band on the 'Y' axis with a channel at 13 GHz.

CA is a future-proof technology which will be utilized more and more to deliver spectrally efficient solutions (Figure 7.20). Specifically:

- mmWave combinations will enlarge to higher spectrum (D-band), allowing us to boost capacity in urban environments up to 1 km, to extremely high capacity

Figure 7.18 Microwave CA deployment example.

Figure 7.19 Carrier aggregation link extension capability.

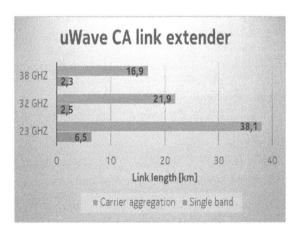

- more and more combinations will interest the microwave + mmWave case – technology enhancements will improve capacity and link distance to exceed the 10 km threshold
- in terms of traditional microwave bands, new coupling will allow stronger integration and optimization, enabling operators to have an effective solution in a smaller form factor

In terms of packet processing, the most effective CA solutions – that fully respond to the capacity and coverage requirements – are based on smart-link aggregation and packet distribution grouping different channels into a single traffic pipe. The specific characteristics of wireless propagation need to be taken into account when designing such systems: as an example, a fast propagation impairment event should not impact high-priority traffic (hitless mode).

The programmability required by future transport networks will need enhanced algorithms, network slicing driven, where traffic steering will depend on KPIs, orchestrated in an SDN environment. CA solutions will need to satisfy additional requirements, imposed by the different 5G use cases in private and public networks (including latency, security and availability) that must be contemporarily provided by the single traffic pipe for different services.

A real network study example shows an existing 2G/3G/4G transport network for a Tier 1 CSP, where today backhaul relies on microwave for almost 90% of sites. In Figure 7.21, cell sites are represented by yellow circles.

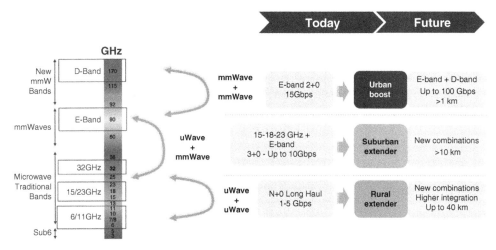

Figure 7.20 Carrier aggregation evolution.

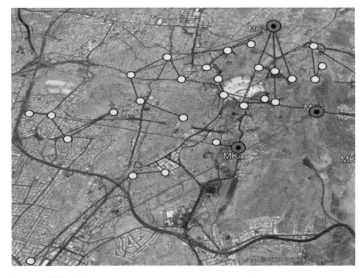

Figure 7.21 Microwave deployment connecting fibre points of presence.

Adding 5G on those sites will increase the transport requirements, which can be fully satisfied with the above-mentioned techniques of E-band (blue lines connectivity) and CA (red lines connectivity) [1].

7.6.3 New Spectrum Above 100 GHz

The telecommunications industry has already started to consider the possible use of frequency bands above 100 GHz for the transport segment of the network. Recent activities indicate the highest interest in D-band (130–174.8 GHz) and W band (92–114.25 GHz) (Figure 7.22).

While W band is viewed today as a likely extension of E-band (71–86 GHz) because of its similar propagation behaviour, the peculiarities of D-band will enable innovative approaches. D-band represents a wireless connectivity solution able to scale up to 100 Gbps, exploiting a 130–175 GHz spectrum, where tens of gigahertz are available thus allowing:

- Fibre-like capacity (N × 25 Gbps) at ultra-low latency (sub-10 μs)
- Ultra-small form factor (antenna size can be down to a few centimetres square)
- More integration between radio and antenna, but also between x-haul and access solution (e.g. small cell)
- Point-to-multipoint and mesh connectivity thanks to beam steering

This shift towards higher-spectrum bands is perfectly in line with the capacity requirements, due to large and unused spectrum, and with the densification need, as transport links will be shorter when an ultra-dense network is deployed. Mesh connectivity will then guarantee the resiliency level which is mandatory for specific SLAs, including mission and business-critical applications.

The very small D-band antenna size can also be exploited to provide a flexible Frequency Division Duplexer (fFDD). This method can be implemented using a double antenna at each end, where one part is reserved for TX and one for RX, avoiding the duplexer filter needed for the traditional FDD systems and enabling stronger flexibility on the customer side because there are no longer radio parts depending on the specific channels used, as in traditional microwave.

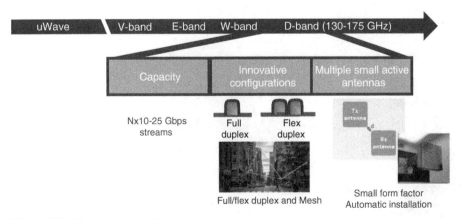

Figure 7.22 Spectrum evolution.

This solution mainly targets dense urban densification associated with the following 5G use cases (see Figure 7.23):

- 5G mmWave deployment (e.g. 26–28–39 GHz) – the first operators starting to introduce 5G through these bands are already considering wireless to complement fibre, as a significant number of cell sites will lack proper fibre infrastructure.
- Remote RRH/small cells/massive MIMO (Multiple Input Multiple Output) deployments – in this case it would be more a point-to-point Ethernet fronthaul connectivity to reach the macro cell or the first fibre point of presence.
- Small cells at street or street/rooftop level – here small form factor and rollout automation are key elements.

Such evolution is possible thanks to breakthrough advancements in silicon design, as well as antenna parts. Silicon technologies are already enabling integrated high-capacity RF solutions, combining in a single chip and package multiple carriers with the required performance in terms of main KPIs (transmit power, sensitivity, noise level, power consumption) – see Figure 7.24.

Moreover, innovative phased-array antennas are under development in order to address installation and operations benefits (e.g. automatic link alignment) and new scenarios like mesh x-haul.

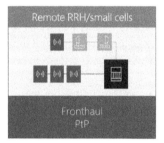
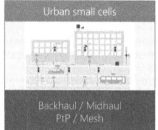

Figure 7.23 D-band use cases.

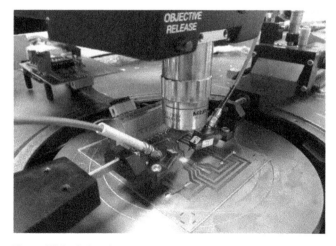

Figure 7.24 D-band technology.

Especially at those high-frequency bands, several innovations are currently under investigation for future market introduction.

- MIMO/OAM

At high bands, geometrical 4×4 MIMO or OAM (Orbital Angular Momentum) techniques are much more applicable, as the optimal distance between the different antennas becomes manageable (in the order of a few tens of centimetres) for actual deployment.

- Beam steering

New techniques like antenna beam steering and self-alignment will allow us to simplify operations and enable full mesh connectivity, which will be necessary when deep 5G densification occur.

- Full duplex

A doubling of spectral efficiency (thus using the same frequency channel to transmit and receive) could be achieved through deep cancellation techniques.

7.7 Radio Links Conclusions

The evolution of the transport network will only be partly covered by increased fibre penetration. Microwave technology will still play an important role in complementing fibre, where the latter is not available (e.g. suburban/rural areas) or not economically viable, especially considering the deep densification trends. Innovative technology advancements in different areas (radio, spectrum, networking) are allowing wireless transport to meet 5G requirements and evolve.

7.8 Free-Space Optics

7.8.1 Introduction

Free-Space Optics (FSO) is a wireless data transport technology like Microwave Radio (MWR). Instead of using radio waves at gigahertz frequency range, FSO uses near-infrared radiation at terahertz range. The theoretical bandwidth is therefore very large compared to radio technology. The wavelength range is typically from 850 to 1550 nm, which coincides with most of the wavelength range used in fibre optics.

Optical frequencies are licence free. Interference with other optical links can be avoided because the beams are narrow, typically less than half a degree. Self-evidently, FSO requires line of sight between the two ends of a link. Thus, a bird flying across the path may block the transmission for a split second. Rain attenuates light moderately but fog is the main limiting factor in achievable link span, and dominates even heavy snow fall. In arid and semi-arid areas, dust storms also affect the link length. Severe sandstorms block any optical links, but these events are very rare even in the areas most prone to such storms. Since fog is the leading limiting factor, it alone will be discussed in the link budget discussion below. Having said that, during snowfall combined with wind towards the optical window, snow could cover the window for hours unless special measures are taken, such as using heaters.

Another limiting factor is eye safety. A safe value for radiation reaching the eye for an extended time at 850-nm FSO wavelengths is 0.78 mW [2], twice as much as the allowed value for visible light. Fully eye-safe operation corresponds to Class 1 laser products. However, the power level is too low for long enough spans. Class 1M, on the other hand, defines that the power can be larger as long as the amount of light that can reach the naked eye is less than 0.78 mW. Since the maximum aperture of the human eye is 7 mm, corresponding to 0.39 cm^2, an irradiance of 2 mW/cm^2 can be allowed. A typical aperture of an FSO device is a few tens of square centimetres. Therefore, optical power in excess of 10 mW can be used while the maximum irradiance remains below 2 mW/cm^2 in all parts of the beam. Even though Class 1M products are safe for the naked eye, they are not safe for observing (e.g. using binoculars) if the observer is in the beam. Therefore, it is necessary to block public access to the beams. The beams are very narrow, typically less than a metre at the receiving end, so eye safety can be ensured without complicated structures. The Class 1M limit for 1550-nm radiation is nearly 26-fold compared to visible light, 26 mW/cm^2, clearly allowing a higher power level. As a drawback, systems using high-power lasers are more expensive and consume more power.

7.8.2 Power Budget Calculations

As in any transmission and receiving technology, the power budget is essential. Power is lost due to geometric losses and attenuation due to fog, for example.

7.8.3 Geometric Loss

Laser beams could be collimated, traversing long distances with minimal divergence and consequently minimal power loss. However, already a slight change in orientation of the transmitted beam would move it out of the reach of the receiver. Therefore, the beams are usually non-collimated, diverging slightly and consequently covering a larger area at the receiving end. If the system employs active alignment, a divergence of 2 mrad (0.1°) or even less is typically used. On the other hand, if active alignment is left out to reduce costs, the divergence is typically 3 mrad or more. Larger divergence causes more geometrical loss, leaving less budget for atmospheric attenuation. To calculate the geometric power loss, first the beam diameter DBeam at the receiving end at distance L is determined using the divergence angle θ. At practical link spans the beam could diverge just a few times the width of the exit aperture. Therefore, the receiver may be in the near field, and the far-field equation [3]

$$D_{Beam} = L \cdot \theta \qquad (7.1)$$

is not necessarily accurate enough. The near-field beam diameter calculations are complicated so we take a simple engineering approach to reach adequate accuracy. In this case we assume that the beam starts to diverge at the divergence angle immediately upon exiting the transmit aperture, see Figure 7.25. Thus, we approximate the diameter of the beam as

$$D_{Beam} = D_T + L \cdot \theta \qquad (7.2)$$

where DT is the transmit optics diameter, representing the beam diameter when it exits the lens.

The attenuation caused by beam spreading can be calculated as

$$A_{\text{beam spreading}}(L) = -10 \lg\left[\left(\frac{D_R}{D_{Beam}}\right)^2\right] = -10 \lg\left[\left(\frac{D_R}{D_T + L \cdot \theta}\right)^2\right] \text{ (dB)} \qquad (7.3)$$

where DR is the diameter of the receiver optics.

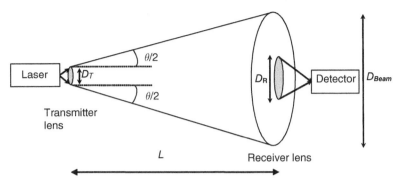

Figure 7.25 FSO link beam spreading.

The equation above assumes that all the power of the beam is evenly distributed within the beam diameters, DT and $DBeam$. This is a crude approximation because the beam intensity profile, especially further away from the transmitter, is approximately Gaussian. The intensity of a Gaussian beam starts to drop outside the centre of the beam and steepens thereafter. The beam diameter is defined as the distance between the points where the intensity has dropped to $1/e^2$ of the intensity in the middle and 14% of the power resides outside the beam diameter. However, the geometric loss calculation is accurate enough to obtain a fair estimate of the fade margin available for atmospheric attenuation.

7.8.4 Atmospheric Attenuation

As mentioned before, only fog is considered because it causes the highest levels of attenuation, although rain and especially snowfall attenuate as well. The attenuation can be described using the formula

$$A_{fog}(\gamma, L) = 10 \lg(e^{\gamma \cdot L}) \quad \text{(dB)} \tag{7.4}$$

where γ is the extinction constant and L is the link span.

To estimate the probability of high atmospheric attenuation in different regions, one can use visibility information gathered by weather stations. However, visibility is defined at visible wavelengths 0.4–0.7 μm, whereas FSO links use infrared radiation. The extinction constant due to scattering by fog is somewhat wavelength dependent. However, it has been difficult to establish agreement about the dependency in the literature. Some models estimate increasing attenuation as a function of wavelength and some decreasing attenuation. Also the droplet size in fog may vary, which varies the wavelength dependence. Therefore, and for simplicity, the model of Ref. [4] is used where the extinction constant is wavelength independent at visible and near-infrared wavelengths below the visibility of 0.5 km.

By using Kim's wavelength-independent variant [4] of Kruse's model [5], the extinction constant can be computed as a function of visibility V:

$$\gamma = \frac{1}{V} \cdot \ln\left(\frac{1}{\varepsilon}\right). \tag{7.5}$$

The visibility is defined as the greatest distance at which a black object may be recognized during daytime against the horizon sky. The definition is based on the contrast ε that is required to remain for recognition. Weather stations used by meteorological institutes and airports (i.e. most weather stations in the world) use the contrast value $\varepsilon = 0.05$. By combining Equations (7.4) and (7.5), and by substituting ε with the numerical value, one obtains

$$A_{fog} = \frac{13.01}{V} \cdot L \ (\text{dB}). \tag{7.6}$$

The total attenuation (omitting inherent losses inside the devices and the minimal scattering due to phenomena other than fog) is

$$A = A_{\text{beam spreading}}(L) + A_{fog}(V,L). \tag{7.7}$$

By taking into account the fixed inherent losses, one can obtain the total system margin.

7.8.5 Estimating Practical Link Spans

Figure 7.26 depicts the combined attenuation of the geometric and fog attenuation of an FSO link at various visibility levels. During clear weather, only the geometric loss remains.

The visibility information of two weather stations was analysed to estimate the availability of two 1-Gbps FSO links depending on link span. One weather station is in a suburb of Hämeenlinna, an average-size town in Finland. The other is in a large park in the heart

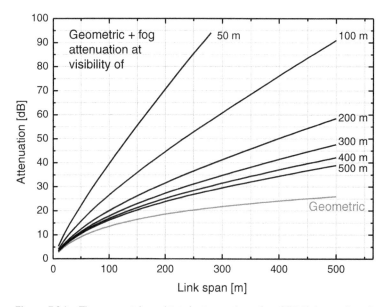

Figure 7.26 The geometric and total attenuation of an FSO link as a function of link span and visibility. The divergence is 3 mrad and the transmit and receive lens diameters are 8 cm.

of Helsinki, the Finnish capital. It can be seen in Figure 7.27 that in Hämeenlinna, the visibility drops occasionally down to 120 m and to 240 m in Helsinki.

The specifications of two 1-Gbps FSO link models were compared with the visibility data of the two weather stations over a duration of 1 year, see Figures 7.27 and 7.28. One FSO system was based on a lower-cost model (less than US$10,000 per link pair) using a single 12-mW, 850-nm beam. The divergence is 2.8 mrad and the beam direction is not steered automatically. The attenuator used to limit optical power against overload remains fixed after installation, limiting the maximum fog attenuation to the dynamic range of the receiver electronics, 19 dB. The beam diameter, after traversing 200 m, is 60 cm. Thus, already a misalignment of 30 cm would reduce the received power by several decibels. Consequently, rigid mechanics and careful installation are required for reliable operation.

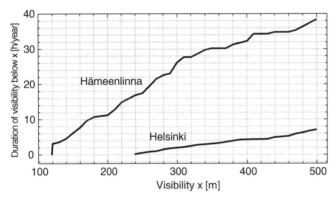

Figure 7.27 The cumulative time when visibility is below a certain distance. For example, in Hämeenlinna the visibility was below 300 m for 26 h during the year.

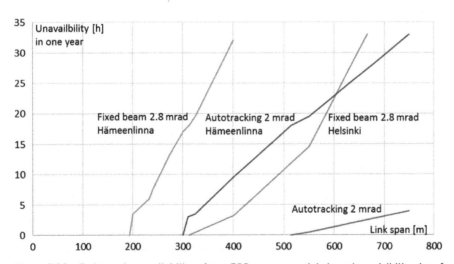

Figure 7.28 Estimated unavailability of two FSO system models based on visibility data from two different weather stations.

The second FSO system uses four 6-mW beams and four receiving optics. One link pair costs more than US$20,000. The beam divergence is only 2 mrad but the link heads have automatic beam tracking that covers an angle of 28 mrad. Therefore, the tracking error remains small even if the tilt or position of the installation platform varies significantly due to, for example, large temperature variation or subsidence of the foundation. It also has an automatic power control utilizing a 20-dB variable optical attenuator. Thus, the dynamic range of the system can be doubled compared to the system described previously. This is essential for withstanding dense fog. For example, in the case of Hämeenlinna, the zero-hour unavailability link span of the lower-cost system could not be improved from 194 m by improving optical performance unless the variable attenuator is also added because almost the whole dynamic range of the receiving electronics is already in use.

If the same FSO links are used for small cells that are used for increasing capacity, and failure of small cells in an area would only lessen throughput but not interrupt service, then 99.95% availability (4.4 h unavailability per year) would likely be acceptable. Thus, based on the 1-year visibility data, for example, the single-beam system could support 200-m spans in Hämeenlinna and 400-m spans in Helsinki.

If the system can drop bit rate during heavy fog and consequently improve sensitivity, then the period of complete unavailability could be reduced. Another means is to incorporate a low-cost lower-bandwidth mmW radio to obtain full availability, since radios can withstand dense fog. On the other hand, FSO withstands heavy rain that could block millimetre waves.

7.8.6 Prospects of FSO

Currently, FSO systems cost more than corresponding microwave systems and provide adequate availability only for short connections. Therefore, they are rarely seen in mobile networks. However, there is potential to decrease the cost of low-power optical links enough to become deployed for small cell backhaul, for example. Bit rates of 1–25 Gbps can be considered today, and higher in the future. Maximum link spans of 200-400 m are feasible, depending on location. FSO is at its best in street canyons when installed above objects that might obstruct visibility but lower than the rooftops where fog is typically denser. At rooftop level the feasible link span drops to about half.

If the bit rate is increased, the receiver sensitivity is usually reduced correspondingly. Therefore, if the distance needs to be increased without dropping bit rate or the bit rate needs to be increased without reducing distance, then one could consider high-power 1550-nm systems since much more power density is allowed for eye-safe Class 1M operation. In the more distant future, coherent detection could be used to increase receiver sensitivity, and consequently bandwidth and distance.

Note that the link span cannot be increased significantly by increasing power. For example, if a system designed to withstand visibility down to 200 m at a span of 400 m were upgraded to support a 500-m span by using a higher-power laser, then the power would need to become almost sevenfold – of which most would be due to atmospheric attenuation.

References

1 Microwave Transport Evolution: An update (2001), *Nokia white paper*, https://onestore.
nokia.com/asset/210908.

2 IEC 60825-1:2014 Safety of laser products – Part 1: Equipment classification and
requirements.

3 Bloom, S., Korevaar, E., Schuster, J. and Willebrand, H. (2003). *Journal of Optical
Networking* 2 (6): 178–200.

4 Kim, I.I., McArthur, B. and Korevaar, E. (2001). Comparison of laser beam propagation at
785 nm and 1550 nm in fog and haze for optical wireless communications. *Proceedings SPIE
4214, Optical Wireless Communications III.*

5 Kruse, P. et al. (1962). *Elements of Infrared Technology*. Wiley: New York.

8

Networking Services and Technologies

Akash Dutta and Esa Metsälä

8.1 Cloud Technologies

In the mobile telecommunications industry, especially in 5G networks, there is increasing interest in deploying the cloud. The key driving factors from an operator's perspective are cost savings, business agility and faster innovation. By deploying network services on commercially available off-the-shelf (COTS) IT servers, operators are looking to minimize inventory and support costs by having to maintain only a few types of hardware (HW) components as against tens or hundreds of HW components in traditional solutions. Deployment of network functions on the cloud also greatly improves new feature deployment via software (SW)-only delivery of features with minimal or no HW impacts. With the advent of the cloud, network function deployment is changing from proprietary box-based network function elements to SW-defined virtual network function elements that can be deployed much more flexibly and with fast-adapting capacity dimensioning as per traffic needs. A data centre and virtualization software combined, referred to here as the cloud infrastructure, is a key enabler in this regard. Let's have a brief look at this next.

8.1.1 Data Centre and Cloud Infrastructure

A data centre is a facility with a pool of physical computing and storage resources networked together to provide a platform to host software services for one or more business applications. The term data centre 'facility' here encompasses everything – the building, the power supply, cooling, physical security and of course the servers and switches therein. As a background, the data centre is an old concept dating back to the days of mainframe computers and computer rooms with large racks of magnetic storage media, but today the term 'data centre' has become synonymous with software-defined data centres with server virtualization infrastructure enabling flexible deployment of applications and services on the data centre. Such data centres form the basic enabler for what we know as the cloud.

The data centre can be operated as a private or public enterprise, depending on the business and security needs. Security establishments usually have strictly private data centres running specific applications with very limited access to the outside world, while players like Google and Amazon operate public or 'internet' data centres, and enterprises or 'tenants' can lease enterprise data centres for their internal IT needs like intranet,

5G Backhaul and Fronthaul, First Edition. Edited by Esa Markus Metsälä and Juha T. T. Salmelin.

employee services and to host their public domain websites. Before going into the specifics of data centres for the Telco cloud, let's quickly look at some general concepts of data centres and the cloud.

The lowest tier in the cloud infrastructure is the HW equipment – the servers and switches in the racks. Typically, an IT/internet data centre consists of General Purpose Processor (GPP)-based servers that fit into standard racks accommodating multiple tens of servers, and there can be multiple such racks leading to hundred or thousand-strong server farms. These 'centralized' or big data centres have robust cooling and power supply systems and therefore the heat dissipation and power consumption of the servers can be higher, to gain from increased compute/storage and networking power. In the Telco cloud, especially due to Radio Access Network (RAN) cloudification in 5G, as we shall see in Chapter 9, there are also smaller data centres, or so-called edge/far-edge data centres, that are deployed in small regional centres or at cell aggregation sites – or even cell sites – and typically have HW accelerators to support NR L1 functions (see Figure 8.1).

This brings some unique requirements to the cloud infrastructure for such deployment sites, mainly in terms of budget for power consumption, heat dissipation and footprint, while satisfying the low-latency, high-throughput requirements for the RAN applications hosted on such data centres. It also brings some security challenges, not only in terms of digital-level vulnerability, but also the risk of leaving unattended multi-thousand dollars' worth of generic IT equipment at such edge/far-edge sites. Nevertheless, there is good momentum towards such edge/far-edge 'mini' data centres, especially in the USA and some other developed countries where operators are starting to see the long-term benefits of cloudifying the 5G radio network functions.

The next higher tier in the cloud infrastructure is the virtualization software infrastructure. There are a few options for virtualization technology, each with their own characteristics. The first to be introduced is what is called hypervisor, which enables us to create virtual machines – literally creating a virtualized machine (or virtualized server) on top of the physical machine. In Telco applications, especially in the RAN, mainly the

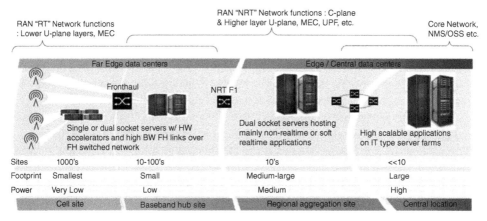

Figure 8.1 Data centre and network function deployment for the Telco cloud.

type-1 hypervisor (especially the Kernel-based Virtual Machine, KVM) is the default hypervisor used, for reasons of performance and ecosystem support. Later, we will see how ETSI Network Function Virtualization (NFV) defines the control and management aspects of this virtualization infrastructure. The virtual machine gets its own virtualized Central Processing Units (CPUs), storage and network interfaces and operating system (Guest OS) running on these virtualized resources provided by the hypervisor running in the Host OS context. This model offers very good isolation of traffic loads and security, taking away a bit of the performance due to the Guest OS and hypervisor layer getting in the way of accessing HW resources. With time, the virtualization technology is moving towards container-based virtualization, where the applications running inside the containers run directly on the Host OS, thus removing the overheads of the hypervisor. In the next section we will look at NFV deployment models utilizing the infrastructure virtualization capability.

Before going deeper into the cloud infrastructure, let's look briefly at cloud service provision models, meaning the level of assistance the virtualized applications can get from the underlying cloud infrastructure. The common service models for the cloud are IaaS (Infrastructure as a Service), PaaS (Platform as a Service) and SaaS (Software as a Service), with increasing levels of abstraction in the service provided from the cloud infrastructure. IaaS such as Amazon AWS, Microsoft Azure, Google Compute Engine, and so on provide a basic capacity of virtualized computing, networking and storage. The clients or tenants have full control over the SW environment of their hosted services and just rent the virtual resource capacity from the IaaS provider. The application scaling and management is developed by the client (or the virtual function owner) on top of the low-level cloud infrastructure provided as IaaS. With IaaS, the pricing is mainly around computing/storage resources, availability level, and so on. The next level of abstraction is PaaS, which includes a middleware framework that completely abstracts the usage of the cloud infrastructure, so users can just focus on the business logic of the applications and offload functions like load balancing, service discovery and messaging framework, scaling, monitoring, and so on to the middleware. This enables faster development and deployment of applications by PaaS users, in turn facilitating faster time to market. Some examples of PaaS are the Google app engine and Amazon Elastic Cloud. However, while PaaS is convenient and faster for service development and deployment, it takes away some control from the application developer in the software runtime environment, as the key here is the re-use of those common/open-source PaaS components. The third model is SaaS, where the application (or application suite) itself is provided by the provider and the client simply uses it for its business needs. Popular examples are Gmail, Cisco Webex and Microsoft Web Office.

Coming back to the HW tier, we now see what a typical data centre looks like in terms of HW architecture and see how the servers talk to each other and ultimately present the group of servers as a pooled computing (and storage) resource that can communicate with the outside world seamlessly (Figure 8.2).

In Figure 8.2, the server architecture is greatly simplified to emphasize the key components there. The dual socket server is the one with two CPUs and typically two NICs (Network Interface Cards). Some of the form factors for dual socket servers are the

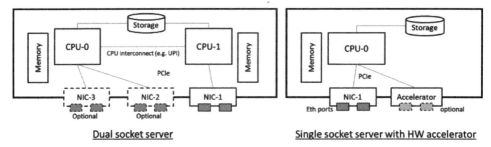

Figure 8.2 Simplified representation of a server.

standard 19″ 1U-high form factor or the OCP-based (Open Compute Project) 19/3″ 1U high, or some specific variation like the 2U-high server for each of those widths. The single socket server is more of a far-edge kind of deployment hardware, needing a smaller footprint and less granular deployment. For the RAN use case, there are now smart NICs, ASIC based HW accelerators and field-programmable gate arrays (FPGA) trending, that help offload functions from vSwitch, cryptography or 4G/5G radio interface protocol functions. The GPPs are becoming more capable – for example, specific instruction set to aid some computation-heavy functions from the 4G/5G radio interface L1, L2 or crypto functions. Customized GPUs (Graphical Processing Units)and DPU (Data Processing Unit) style accelerators are being considered for hosting the real-time radio L1 virtualized functions in the cloud. So what was seen as not feasible to be done in the cloud, typically in L2 and L1 RAN functions, is now becoming a reality, and starting to promise big benefits to Telco operators. In addition to the 3GPP standard protocol functions, AI/ML (Artificial Intelligence/Machine Learning) augmentations are fast becoming integrated to the RAN network function elements like vDU, RU and even vCU.

As we said at the beginning of this chapter, the core of the data centre is this pool of tens or hundreds or thousands of servers, each with one or two CPUs and storage/memory, all cabled together via switches and routers to form the cloud, where the applications or network functions can be hosted to make use of this pool of virtualized resources in a use-as-needed manner.

8.1.2 Data Centre Networking

Coming to the networking aspects, as said above, a data centre is a pool of servers connected via switches and routers. In the central office deployment, there are typically several racks housing multiple servers, each along with rack-local (so-called Top-of-Rack, ToR) switches. There are multiple ways in which the Data Centre Network (DCN) can be designed, like the three-tier DCN, Clos topology variants like the fat tree, DCell, spine–leaf, and so on. Let's take a quick look at the most popular DCN architectures.

The main challenges for DCN design are scalability, resiliency, efficient bandwidth utilization for the interconnect links and cost as a whole. In the cloud, there is usually enormous east–west traffic (inter-server traffic, within the data center) in addition to the natural north–south(in/out of/from the data center) traffic. For example, a Google search can take thousands of inter-server interactions before presenting the results to the user – these servers can be in the same rack or different racks, or even different locations. All this inter-server traffic is within the cloud and data centre (simplified example, not limiting), and it can be 10× the actual traffic to/from the end user (north–south, with simplification).

For years, the cloud DC network has been designed based on the traditional three-tier DCN with three levels of networking – access/edge, aggregation and core. There are servers connected to the ToR switch (access) that are in turn connected to the aggregation switches that are connected to the core routers. Typically, there is an L2 network between servers and access switches and up to aggregation, and upwards there is L3 routing. There is full redundancy in every layer to eliminate single points of failure in the traffic path.

Today, the spine–leaf Clos fabric is the default topology. It has only two layers – leaf switches that connect to the servers and spine switches that connect to the spines (see Figure 8.3). All leaf switches are connected to all spine switches, so all inter-rack servers are the same distance (or hop) apart, with full mesh connectivity. Equal-cost multi-path routing can be used to distribute the traffic to the spines over all links, and gives excellent performance for east–west traffic that is the main driver for this architecture.

For the Telco cloud there are some specifics to the cloud DC needs and architecture. First, the Telco cloud is operated like a private cloud, typically owned by the Telco operator

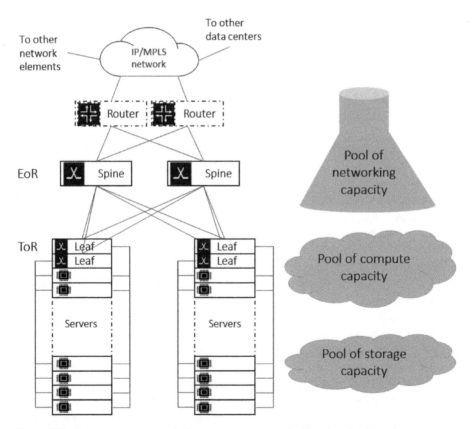

Figure 8.3 Data centre as a pool of resources connected with spine–leaf topology.

and the end users are not accessing it from the web browsers (like Facebook or Google). Note that Telco network function deployment on a public cloud is a possibility, but a few years away from now. Second, the Telco cloud, or more specifically the RAN cloud, has specific applications deployed, providing the so-called NaaS (Network as a Service), and a key characteristic is that such a service is less data driven (unlike Facebook or Google Search kind of services) and more about providing a set of protocols that talk to the user terminal to give communications access. Additionally, the Telco functions deployed in a DC are generally for a relatively localized geographical area (a region or large city and its suburbs). As a result, the Telco cloud need not be hugely scalable, has less east–west traffic in general (especially in the case of cloud RAN) and may have some specifics like HW accelerators and special plugin boards. There can be some soft real-time and even hard real-time applications in the cloud, depending on the choice of cloud RAN deployment. We will see these deployment options further in Chapter 9.

8.1.3 Network Function Virtualization

NFV is the concept of realizing one or multiple network functions as virtualized building blocks that can be combined or chained together to provide the targeted communications services. Each Virtual Network Function (VNF) can be realized in software and deployed on standard IT servers in the virtualized infrastructure to provide a certain self-contained network function, such as a load balancer, router, radio network controller, and so on. The NFV is an essential enabler for the Telco cloud. One of the main motivations is to optimize the cost of buying and maintaining several tens or hundreds of proprietary HW and SW components for the Telco services. NFV enables deploying most of the network functions (as VNFs) on standard IT servers, with the added benefit of elasticity to better cope with the potentially rapidly fluctuating demand of network traffic. What this also enables is that the Telco operator can lease a virtualized data centre and host variety of applications and services, ranging from billing to management, orchestration, core network functions, parts of RAN functions, MEC, content servers, its own intranet, and so on – all on the same virtualized infrastructure.

The NFV framework is formalized by the ETSI Industry Specification Group (ISG) [1] with a full suite of specifications and guidelines for virtualization infrastructure, management and orchestration and the VNFs. Figure 8.4 shows the ETSI NVF reference architecture.

At the high level there are three working domains in the NVF architecture: the VNFs (application domain), the NFVI (infrastructure domain) and the M&) (Management and Orchestration) domain.

The VNF is the (virtualized) network function, that was in the legacy world realized as a Physical Network Function (PNF). Telco VNFs can be part or all of the legacy network element, typically a whole network element such as LTE MME, LTE SGW, 5G CU-C/CU-U, MEC, a security gateway, and so on. A VNF is composed of multiple components, or VNFCs, that can be deployed as individual Virtual Machines (VMs) or pods. The EM and OSS in the application domain are for business-level management of the VNFs and can also interlink with the NFV management and orchestration to synchronize the business and infrastructure management aspects.

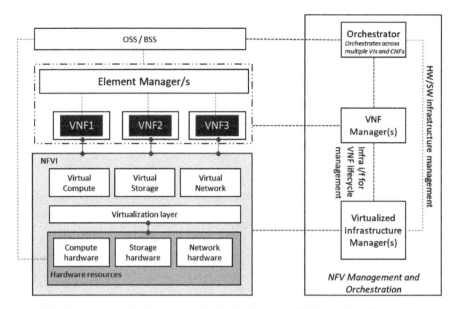

NOTE: this is a functional split, not necessarily depicting location split of the functional blocks. For example, VIM is typically embedded in the Virtualization Infrastructure

Figure 8.4 ETSI NFV model.

The NFV Infrastructure (NFVI) are the HW resources and software framework for virtualization of computing, storage and networking, the aspect we covered in the previous section. In other words, the switches and the host operating system (including the hypervisor or container engine) are all part of the NFVI. The VNFs are deployed on this virtualized infrastructure.

The management and orchestration domain are key to providing the capabilities and benefits of the cloud. At the core of the NFV management is the VIM (Virtualized Infrastructure Manager). The VIM is the manager of the NVFI instance and manages virtualized resources, allocates and reclaims the virtualized resources to the VNFs and/or VNFCs, and provides performance monitoring and fault management of the NFVI. Open Stack is a good example of a VIM. The VNF manager is the entity responsible for lifecycle management (instantiation, update, scaling, etc.) of the VNF via the VIM and/or directly with the VNF. The NFV orchestrator is the top-level management for overall management and automation of infrastructure and network function deployment. It is noteworthy that VNFM, NFVO and EM can themselves be VNFs.

As an example, a Telco operator may have a cloud LTE network with the MME, and other core elements fully virtualized as VNFs and deployed in its central data centre. The same place can also host some of the RAN elements as well as the OSS, billing a kind of virtualized functions. So, in effect, large pools of HW resources, abstracted by the virtualization framework and managed homogenously via the cloud management and orchestration function. So, what the NFV enables is this transition from the traditional proprietary platform-based network elements to the software-defined network elements, and

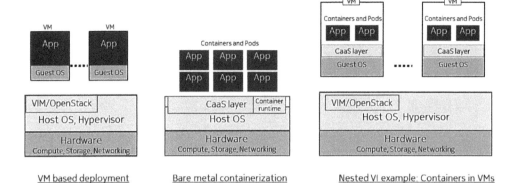

Figure 8.5 Virtualization options.

this in turn opens a whole new world of opportunities to play with extremely flexible deployment, automated management and on-demand scaling with very high reliability for all VNFs.

Figure 8.5 shows a simplified illustration of virtualization options to deploy network functions. The latter can be deployed as a set of VNFCs as VMs or pods (which may have one or more containers). We will look a bit further into containers and pods in a while. Apart from the two options mentioned above, there is also the possibility to have a nested virtualization – for example, applications running inside containers deployed inside VMs. There are more virtualization nesting combinations possible, but they are beyond the scope of this book.

8.1.4 Virtual Machines and Containers

VMs are the most isolated mode of virtualization, creating a virtually independent machine (or computer) inside a physical server or computer, thus virtualizing at HW level. There is a slight performance and memory footprint penalty with VMs due to the presence of the hypervisor layer and two separate operating systems – the Host OS on the physical server and the Guest OS running inside the VM. This penalty is avoided by the other common virtualization technique, called containerization. Containers provide lightweight virtualization by running directly on the Host OS, with isolation of container runtime provided by Host OS and container runtime capabilities.

Isolation of containers is achieved via namespace separation per container, enabling dedicated/isolated process trees, network stack parts, routing rules and filesystem. A pivotal enabler of container isolation is Linux control groups (cgroups), which allow prioritization and constraint of HW resources (CPU, memory, network, etc.) for a group of processes so that a container cannot infinitely hog any system resources and cannot get into the resources of other containers. The Google-driven containers orchestration framework – Kubernetes – is the de facto containers management and orchestration framework and is a graduated CNCF (Cloud Native Computing Foundation) project. Kubernetes, a.k.a. K8S, introduces a deployment abstraction (on top of containers) called the pod. The pod is the minimum deployment unit in K8S and can hold one or more containers. As a simplified reference, the equivalent of VNF in the containers/pod world is the so-called

CNF (Cloud Native Network Function). A CNF typically has multiple pods having containerized application software deployed in those pods. A key expectation from a containerized application is that it should cater to a small capacity and be composed ideally of a single process, cloud native in the sense that it is stateless and developed with microservice design pattern. The additional advantage of container-based virtualization – apart from resource efficiency – is fast lifecycle operations (e.g. create, delete, scale, etc.), which in turn allow much faster recovery from failures and fine-grained scalability.

The following paragraphs give a very brief overview of some networking options for virtual machines and containers.

8.1.4.1 Virtual Machines and Networking

Although a lot happens under the hood in VM or containers/K8S runtime, we restrict ourselves here to high-level coverage of the networking aspects of the VM-based or containers-based deployments. Figure 8.6 is a simplified illustration of networking and communications in an OpenStack/KVM-based cloud infrastructure.

This figure shows a simplified connectivity of VMs on a server and highlights an important networking element – the vSwitch. The vSwitch is simply a virtualized switch (i.e. a switch implemented in SW) with virtual ports connected to VMs and to physical interfaces on the server NIC. So, the setup is just like having multiple physical machines wired to a physical switch. While the built-in Linux bridge does a good job of bridging inter-VM traffic within a node, the vSwitch extends the inter-VM and external connectivity beyond a single node and simplifies the manageability of VMs' connectivity by responding (via the orchestration framework) to the changing network dynamics that is common in cloud systems. The vSwitch (e.g. Open vSwitch, OVS) can also be DPDK-accelerated (Data Plane Development Kit) to handle low-latency, high-throughput use cases. As shown in Figure 8.6, the VMs can have one or more virtual ports connected to the vSwitch. These VM interfaces (vNIC ports) can then be on tenant internal networks or connected to the provider/external networks via the vSwitch. There are also virtual routers available that extend the functionalities.

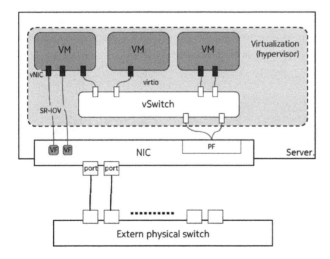

Figure 8.6 Networking solution for virtual machines.

In Figure 8.6 there is another alternative shown – SR-IOV – which is typically used for low-latency, high-throughput traffic (especially if a DPDK-accelerated vSwitch is not available). SR-IOV (Single-Root IO Virtualization) allows a physical NIC to expose multiple virtual functions (VF) that can be accessed by the application (in a VM or container/pod) just like a network interface. Key advantage of SR-IOV is that the virtualization (provision of VFs) is done right at the PCIe level, which eliminated any SW based virtualization overheads, thus giving almost NIC HW level performance to the virtualized application. The SR-IOV-enabled NIC has an embedded bridge for switching traffic across VFs. It is noteworthy that an NIC can have single, dual or more 1/10/25/50/100 + G ports with breakout/lane separation options on its ports (e.g. a QSFP28/100G port could typically be configured as 1×100G or 2×50G or 4×25G or 4×10G lanes). This means that the NIC port can have one or more PFs or MACs, each giving multiple VFs, in turn giving a highly flexible networking solution. Still, a vSwitch or vRouter-based solution is much more flexible and easier to manage compared to an SR-IOV-based network solution. Therefore, SR-IOV-based interfaces are used only where needed, if no other alternatives are available.

After the infrastructure view of the networking solution, let's look at the high-level networking solution from the VNF perspective – the tenant network for the inter-VM/node traffic, or what is called the east–west traffic, and the external/provider networks that give external access to the VMs that carry the north–south traffic. A popular way of separating these networks is by using different VLANs or VxLANs per tenant and provider network. From an E2E connectivity point-of-view, each VNF may have one or more networks attached, with public IP addresses assigned on one or more interfaces of its VMs (or VNFCs). The example in Figure 8.7 exposes the external IP addresses directly to the collaborating peers of the NFV in question. An alternative is to have a transport front end that acts as the gateway in and out of the VNF for all kinds of traffic to the peers. This helps with seamless scaling and recovery of back-end services without impacting the peer/remote transport endpoint(s).

NFV is moving rapidly from VM based to container based. So, after a brief view of VM-based networking, let's have a look at some networking constructs of the containers/Kubernetes world. We will see only the Kubernetes networking model, as that is the default CaaS platform in the industry now.

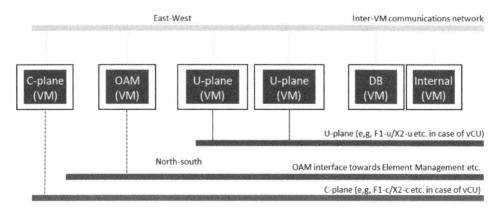

Figure 8.7 Tenant networks with virtual machines.

8.1.4.2 Kubernetes and Networking

As we said earlier, the virtualization technique of choice is fast shifting to containers and Kubernetes, as these offer much lighter virtualization in terms of resources as well as time (to create/scale the virtualized functions). While containers provide the basis for this light-weight virtualization technique, Kubernetes is what enables the management or orchestration of container-based applications. Kubernetes brings a simple and robust framework for deploying the containers via so-called pods and network plumbing to enable them to talk to each other and with the outside world.

The Kubernetes network model has three principles (see Figure 8.8):

- All containers can communicate with all other containers without NAT
- All nodes can communicate with all containers (and vice versa) without NAT
- The IP that a container sees itself as is the same IP that others see it as

This can be achieved in a variety of ways and therefore Kubernetes offloads the actual net-work setup to the network CNI (Container Network Interface) plugin. Some popular choices for the default network (cluster network, or pod net) are Calico, OVN-K8S and Cilium. The default network can also carry inter-pod application control/management traffic, though application traffic can also be carried over a separate network using IPVLAN or SR-IOV, depending on the capabilities and characteristics of the default network. For example, while Flannel uses an overlay network for inter-pod communication, Calico gives the advantage of direct routing-based inter-pod communication, thus avoiding overlay net-work overheads. In many applications – especially vCU or vDU – there is often a need to have more than one network interface per pod. To enable this, a meta CNI like Multus or DANM can be used to attach additional network interfaces as per the need of the

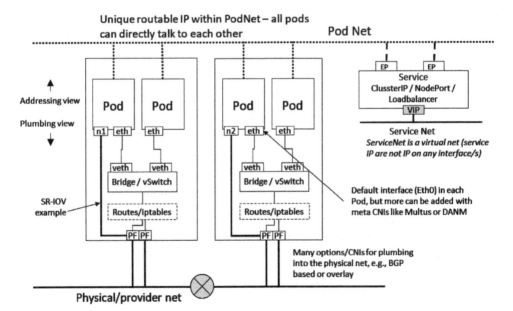

Figure 8.8 Networking in Kubernetes – simplified view.

application. Typical deployment of vCU and vDU would have a default (cluster) network and additional networks over IPVLAN, and particularly the SR-IOV-based interfaces (DPDK or non-DPDK). SR-IOV is specifically useful in RAN network functions for performance reasons, typically for north–south user plane traffic on F1 and fronthaul interfaces in vDU. Control plane and OAM (Operations, Administration and Management) north–south traffic for the Central Unit (CU) and Distributed Unit (DU) can be mapped, for example, to IPVLAN as the throughput and latency jitter is relaxed on these interfaces, though SR-IOV (non-DPDK) also fits.

There are options to either directly expose the application pods to the external network (i.e. network outside of the data centre) or have a layer of a load balancer between the application pods and the outside network. Kubernetes has some useful networking constructs, like Service and Ingress/Egress, that can be used to provide a neat entry/exit point for the application pods, but there are some limitations or assumptions that should be considered. These networking constructs work best when the application is designed as a set of truly N + stateless services, without the need to have any specific stateful load balancing or context maintenance logic in the load balancers. However, for some RAN CU/DU applications, especially the ones implementing user plane protocols on fronthaul, it is challenging to be truly stateless. Also, some 3GPP protocols like SCTP are not well supported by the Kubernetes networking framework (SCTP support for Service/Cluster IP is in Beta in Kubernetes v1.9), so utilizing some of the native Kubernetes capabilities is limited in RAN applications. As an alternative, an application-level solution can be provided for load balancing with RAN application pods directly exposed to the external network.

From a CNF perspective the networking solution in the Kubernetes environment is quite like that shown in Figure 8.7 for VM-based deployment. The mechanics are different, but the logical connectivity and even IP address management are similar in the sense that typically the tenant inter-pod or inter-VM IP addresses are fully managed by the infra (CNI IPAM in case of Kubernetes) and the external IP addresses can be managed either by the infra or by the RAN operator via the NMS or another mechanism (e.g. ConfigMap).

8.1.5 Accelerators for RAN Functions

Broadly, two types of HW accelerators are considered – lookaside and inline. The lookaside accelerator offloads some part of the processing and the result it taken back into the parent application running on the server CPU, which in turn does the actual sending and receiving of the data stream on the external interface. This means there is some additional latency to transfer the data across the (typically PCIe) interface between server CPU and HW accelerator. The need for the HW accelerator is based on TCO (Total Cost of Ownership) analysis based on the traffic use case, for example. It is a balancing act of how much to offload so that there is an overall reasonable balance between CPU compute resources and its sufficient use, while still offloading some functions to the HW accelerator. HW offload should not end up under-utilizing the main CPU for the traffic or deployment scenario. Though the virtualization is applicable to both CU and DU, more specialized support is needed from the cloud infrastructure for DU virtualization as it has more real-time functions with specific processing functions for L1.

Looking at the CU, the main compute needs are towards processing of C-plane protocols, slow RRM (Radio Resource Management) algorithms (e.g. admission control, handover control, RRC connection management) and U-plane non-real-time functions (GTP-u, SDP, PDCP protocols). While the C-plane and related slow RRM functions are truly non-real-time (response latency of millisecond order doesn't cause any issues) and low to moderate bursty traffic, the U-plane is the main consumer of compute resources. Air interface encryption and transport security (IPsec) are the big consumers, and a new one is air interface integrity protection, which can be extremely taxing depending on the choice of the hashing algorithm (AES is simple, while ZUC and Snow are very complex). So, the main candidates for computing offload to a HW accelerator are these security functions. Initially, the security accelerators are lookaside (Figure 8.9) (standalone security function accelerators and smart NICs that include such functions). There are some processors already available today that include an embedded security accelerator on chip.

For vDU, the use of a HW accelerator is almost necessary, mainly for L1 processing. There are multiple cloud infrastructure options for the cloud RAN solution to meet the compute needs of the DU. The amount of offload is based on the HW accelerator capabilities – a lookaside FEC HW accelerator just offloads the FEC to the HW, while the rest of the L1 processing and fronthaul termination happens in the main CPU of the server. An in-line HW accelerator offloads typically almost full L1 and has the fronthaul function and interfaces so that the data can go out directly from the HW accelerator to the RUs (Radio Units) (and vice versa). This is the key difference between lookaside and inline accelerators – an inline accelerator does the processing 'in line' without the need to return data to the server CPU/s for external transmission and reception. This saves the cycles on the server CPU/s (GPP cores) as well as helping latency, but the flip side is the need for an additional HW (accelerator) to offload more from the server CPU. So again, as said before, it is a balancing act between finding the right solution for the target use cases, and the landscape is still evolving in the sense that there is no clear winner yet (especially keeping a long-term view) in the comparison between the two accelerator modes.

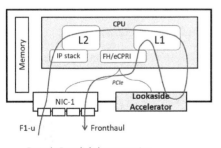

Example Downlink data processing with lookaside L1 accelerator

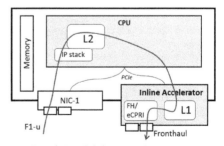

Example Downlink data processing with inline L1 accelerator

Figure 8.9 Lookaside and inline accelerators for gNB physical layer function.

8.1.6 O-RAN View on Virtualization and Cloud Infrastructure

O-RAN (Open RAN) aims to make the RAN true open and multi-vendor. While 3GPP makes interfaces like X2, S1 and so on standardized, it is well known that it does not guarantee that the network functions on each side of an interface (say CU and DU over the F1 interface) can work if they are made by different vendors. This can happen due to incompatible interface termination and interpretation at both endpoints, in terms of supported information elements and the business logic behind the interface. O-RAN sets this interoperability clearly, by clarifying the interface definition where needed and most importantly defining the interface compatibility profiles for the network function vendor to satisfy being O-RAN compliant. O-RAN also defines a new network function – the RIC (RAN Intelligence Controller). The RIC comes in two flavours – near-real-time and non-real-time – and together they can control slow decisions (like policies) and fast decisions (like parameters or outcomes of algorithms) in the RAN, respectively. The idea is to be able to run value-added intelligent functions outside of the RAN network service functions like CU or DU, so that a telecom operator may either source those from cheaper vendors or make their own customizations without dependency on the RAN vendor. The RIC has interfaces (called E1) to O-CU and O-DU. O-RAN also combines Telco management and cloud orchestration functions into the so-called SMO (Service Management and Orchestration) framework. This introduces new management interfaces as well – O1 for Telco management functions and O2 for cloud orchestration functions.

O-RAN covers a variety of topics on the RAN architecture, including inter-network element interfaces, virtualization, deployment, RAN intelligence separation/control and management/orchestration aspects. Here we focus on some selected topics that are interesting from a cloud infrastructure perspective:

- CNF deployment
- cloud infrastructure (more interestingly, the AAL concept)

From a CNF (vO-CU, vO-DU) deployment perspective, the cloud NFs deployment scenarios B and C are the main cloud RAN deployment cases as defined in the O-RAN Cloud Architecture Description (ref O-RAN.WG6.CADS-v04.00). An exemplary scenario is illustrated in Figure 8.10.

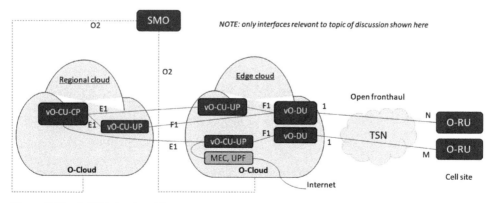

Figure 8.10   O-RAN cloud architecture.

There can be quite a few combinations of how and where vCU-CP, vCU-UP and vDU are placed. Both could be on the edge cloud, or vCU-CP could be more centralized in the regional cloud. These deployment scenarios are not O-RAN specific and will be discussed in Chapter 9 as well. The basic assumption for cloud RAN deployment is that some level of baseband processing centralization is utilized to get gains from pooling and resultant resource and power savings. This leads to an important capability of a cloud RAN solution – the ability to dynamically adjust the O-cloud footprint based on traffic load. This is discussed further in Section 9.2.

O-RAN defines reference cloud platform requirements based on location. For example, the servers in a regional cloud are expected to be standard off-the-shelf servers, while those in edge locations are supposed to be optimized for smaller real estate and power budget. Similarly, regional network/connectivity requirements can be thought of much more in terms of throughput, while for the edge cloud they are more about latency (especially on the fronthaul) and provision of accurate sync for the servers that host vDU. HW accelerators are by default expected to be used in the far-edge cloud (for DU). O-RAN also provides some reference HW and SW profiles for the O-cloud.

8.1.6.1 HW Acceleration Layer

O-RAN has defined the so-called HW Acceleration Abstraction Layer (AAL) that attempts to standardize the interface between the server and the HW accelerator. This includes the data plane interface and the accelerator management interface. Though the use of a HW accelerator is not limited to L1 on the DU, that is the foremost use case of this concept. The central idea is to generalize the interface for the variety of ways HW accelerators can be used by the CNF (O-CU or O-DU in case of RAN), so that cross-vendor operation is possible with a CNF application and HW accelerator from different vendors. Here we focus on the use of a HW accelerator by containerized application, though O-RAN also includes the use of a HW accelerator by VM-based applications.

Figure 8.11 illustrates, without going into detail, the context of the HW accelerator in terms of its interfaces towards the radio application and its management function. In O-RAN, SMO is the control centre for both FCAPS of telecom network functions (like the vCU/vDU, or so-called O-CU and O-DU in O-RAN parlance) and lifecycle management of the CNFs and cloud infrastructure (or the O-cloud in O-RAN terms). SMO has a standard O2 interface towards the IMS (Infrastructure Management Service), which in turn is extended to use the HW accelerator manager. The key goal is that any HW accelerator from any vendor should be manageable through the standard O2 interface from the CMO. So, any vendor-specific interface should be limited to the interface between the HW accelerator and its manager. The HW accelerator manager should support the O-RAN standardized AAL-C (common) management interface towards the IMS, thus abstracting away any vendor-specific application programming interfaces from the IMS.

On the telco application interface, there is a concept of LPU (Logical Processing Unit). The LPU concept allows us to partition the HW accelerator potentially into multiple logical accelerators, both in terms of capacity and function. For example, an L1 HW accelerator could have the capability for one or more inline L1 LPUs and additionally provide acceleration for encryption or beam-forming weight calculation. Such an accelerator could be used by multiple functions in, say, an O-DU, like L1 (complete inline offload), L2 (for

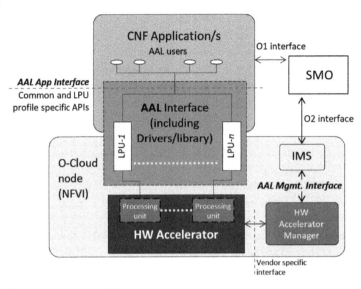

Figure 8.11 O-RAN AAL overview.

beam-forming weight calculation in lookaside mode) and its IP stack (for IPsec in lookaside mode). Of course, the initial accelerators out there are focused on a single function.

It should be noted that during the writing of this book, the O-RAN specifications are still in progress and details of these interfaces are yet to be standardized.

8.2 Arranging Connectivity

8.2.1 IP and MPLS for Connectivity Services

A common way to deliver IP connectivity is with IP/MPLS technologies.

An IP MPLS VPN [2] for mobile backhaul application could be something like shown in Figure 8.12. A key building block in the control plane is multiprotocol-BGP (MP-BGP [3]), which has been extended in several ways to support new services like E-VPN [4].

Further alternatives have emerged with segment routing, which is supported with MPLS data plane (SR-MPLS) and IPv6 data plane (SRv6) [5, 6], which will be discussed separately.

In Figure 8.12, each of the cell sites (gNBs) connects to a hub point multilayer switch/ router (CE) which attaches to the IP VPN service. The terminology defines the CE (Customer Edge) router as the one interfacing the MPLS network edge, the PE (Provider Edge) device. The CE can also be a host device, like gNB.

The MPLS network P (Provider) nodes focus on forwarding based on the MPLS Label Switched Path (LSP) tunnel labels. The CE device is attached to two different PE nodes on two different sites for resiliency. For NG2 and NG3 interfaces the peer is in the core network site with UPF and AMF, and is dual attached as well.

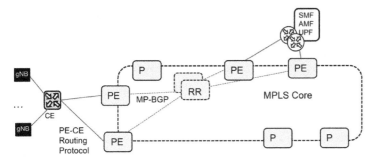

Figure 8.12 IP MPLS VPN.

The intelligence is in the PE devices that run MP-BGP and distribute customer prefixes from and to the CEs and PEs. Between CE and PE, preferably a routing protocol is used; this could be BGP or some other routing protocol, like OSPF or ISIS. Redistribution of information between routing protocols is required in the non-BGP case. For each VPN service, customer routes are kept separate, with the help of VRF (Virtual Routing and Forwading) in the PEs and MP-BGP operation itself.

In the control plane, an MP-BGP route reflector (RR) is used to avoid full mesh of MP-BGP sessions between the PE nodes. The route reflector is duplicated to avoid a single point of failure in the control plane.

In the data plane, MPLS labels are attached to IP packets arriving at the ingress of the PE. In the basic case, two labels are added: one inner label for the VPN service and another outer one for the tunnel. MP-BGP allocates the inner labels, while the tunnel label could be allocated also e.g. by LDP (Label Distribution Protocol).

In the provider MPLS network, an Interior Gateway Protocol (IGP) takes care of the topology related to the LSPs. This is not (directly) visible to the customer networks and vice versa, the P routers do not see the customer prefixes.

This approach has multiple benefits for connectivity service:

- The approach is very scalable
- Different connectivity models can be supported for customer sites, such as any-to-any or hub-and-spoke
- Each site can be dual-homed for resiliency, in an active–active manner for efficient use of network resources
- For 5G network slicing, each network slice can have its own IP VPN, enabling traffic separation
- New customer sites and new customers within existing sites can be added for scalability
- Services can be added (e.g. for traffic engineered paths or fast re-route)
- Connectivity services are supported for both IPv4 and IPv6 and also for Ethernet (with E-VPN)
- Overlapping customer IP address spaces can be supported (different VPNs)

8.2.2 Traffic Engineering with MPLS-TE

With MPLS traffic engineering (MPLS-TE), a path can be reserved through the MPLS network. A signalling protocol reserves resources from all of the nodes along the path, and in this way capacity can be guaranteed. Essentially a circuit (an LSP) is created through the network and this circuit is set up and kept alive by the control protocol, RSVP-TE [7]. This can be combined with the IP VPNs so that one possibility for the tunnel is a Traffic Engineered (TE) tunnel (Figure 8.13).

A link-state routing protocol with TE extensions is required. The head-end node can calculate the path based on this information from the network, or the calculation can be offloaded to a path computation engine. The information from the network is collected into a TE database that is used as input for the calculation. With this information, an LSP is created and resources reserved. For traffic in the reverse direction, another LSP is needed, starting from the node marked 'tail end' in Figure 8.13, which becomes the head end for that direction. A separate PCE element may be useful, especially for complex calculations or for the case where the LSP crosses multiple areas.

TE is important for many use cases, as normally IGP shortest paths are used, which may not be optimal. TE can provide a specific path through the network that does not follow the shortest path, reserve resources for guaranteed bit-rate services, distribute load on the network more evenly and provide protection. A fast re-route capability is a pre-calculated back-up path that is used in case of failure of the active path, reaching sub-50 ms recovery, with a local decision.

As a further step, the MPLS TE control plane, RSVP-TE and IGPs with TE extensions were defined as the G-MPLS control plane [8]. G-MPLS (Generalized MPLS) expands basic LSPs with support for optics, λ-switch-capable LSPs, so that the optical domain is combined with MPLS into a generalized MPLS architecture for MPLS services.

8.2.3 E-VPN

Ethernet VPN (E-VPN) [4] is an evolution from the earlier L2 VPN VPLS (Virtual Private LAN Service) technology and can also be viewed as an adoption of IP MPLS VPN principles to Ethernet services. Improvements with E-VPN for Ethernet services allows a dual-homed attachment to PEs for resiliency and an active–active use of links.

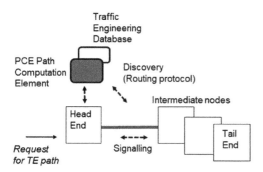

Figure 8.13   MPLS TE architecture example.

E-VPN RFC includes IP VPN services, making it a combined L2 and L3 VPN technology, which is a benefit in use cases where both Ethernet and IP services are required. There is also a possibility to use other than MPLS LSPs as the tunnel layer, with IP/GRE (General Routing Encapsulation) or other methods.

Figure 8.14 illustrates an example where the E-VPN service is used to interconnect data centres with multihoming. The service could be an IP VPN or an Ethernet service, as Ethernet service is now extendable over wide area networks.

E-VPN implementation relies on the MP-BGP control plane, like IP MPLS VPN. While in IP MPLS VPN services customer networks are advertised by the PE–CE routing protocol, in E-VPN Ethernet services customer MAC addresses are learned by the PE. After that, the addresses are distributed by the MP-BGP control plane to the PEs attached to the remote customer sites – as are the customer networks in IP MPLS VPNs. MP-BGP address families have been extended to include MAC addresses (new route types). Route Reflectors (RRs) are used to increase scalability in the control plane. Learning of MAC addresses is limited to the edges and after that, control plane is used. This limits flooding and makes the network scalable.

With Ethernet, one needs to take care of Broadcast and Unknown-unicast/Multicast (BUM) traffic to avoid loops, duplicate frames or MAC address instability. At the same time, it is required to support dual homing of Ethernet segments to different PEs on different sites and have active–active forwarding as with IP MPLS VPNs. With E-VPN, BUM traffic needs to be handled separately. This Ethernet-specific handling adds functionality compared to IP VPNs.

In addition to the Ethernet LAN service for multipoint connectivity, where the E-VPN appears as a bridge, Ethernet point-to-point service (Virtual Private Wire Service, VPWS) is defined in RFC 8214 [9] and Ethernet E-Tree service in RFC 8317 [10]. E-Tree defines a network with root and leaf nodes. The leaf nodes cannot talk to each other, only with the root.

Provide Backbone Bridging (PBB) [11] 'mac-in-mac' can be combined with E-VPN to reduce the number of MAC addresses visible at the edge (PE) to those aggregated behind another B-MAC (Backbone MAC) address. With Ethernet services as well, it is possible with MP-BGP in E-VPNs to flexibly determine which addresses get distributed to which sites.

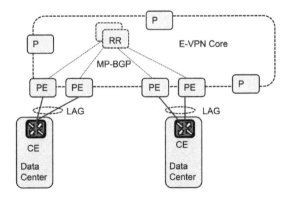

Figure 8.14 E-VPN with multihoming.

8.2.4 Segment Routing

Segment Routing (SR) relies on source routing paradigm instead of hop-by-hop routing. Source nodes include a list of instructions (segments) for the packet that may, for example, be nodes to be visited along the path to the destination (like with source routing); here the segment is a topological instruction. A segment may also be a service instruction that asks for specific local processing or action at the node. A simplified principle is shown in Figure 8.15.

In Figure 8.15, a segment-routed tunnel from Node 1 to Node 6 is defined by a segment list consisting of Node 2, Node 4 and Node 6 (illustrating the SRv6 variant). At the first segment, Node 2 forwards the packet to the next segment, Node 4. Node 4 sees Node 6 as the last entry in the segment list and forwards the packet towards it. At Node 6 the segment list is empty. As mentioned, the segment could also represent an instruction.

There is a variation in how the segment list is constructed, depending on the data plane alternatives and further details therein. In the IETF SPRING (Source Packet Routing in Networking) working group, both MPLS data plane and IPv6 data plane alternatives are defined. With MPLS the segment is encoded into a label stack (RFC 8660), while with IPv6 the segment is encoded as an IPv6 address in the SRv6 header (RFC 8754) [5, 6]. With IPv6 there is no need to deploy MPLS, which is a benefit if there is no prior MPLS deployment in the network. The segments are distributed by control plane (IBGP, OSPF, IS-IS) or populated by the management plane (e.g. by SR controller). The segment list is calculated either locally or centrally by the Path Computation Element (PCE).

SR as mentioned can solve for TE as e.g. in Figure 8.15, where the tunnel does not use the shortest path from Node 1 to Node 6. Differences compared to MPLS TE are that instead of a state (soft state) being created at each of the nodes along the path to realize a circuit, the SR ingress node populates a segment list, after which the packet is forwarded to the ultimate destination segment by segment, with the segment info given in the packet. With SRv6, the next hop does not always need to be defined as the next segment. It is feasible to have nodes along the path that are not encoded in the segment list. In this case the shortest path is selected by the transit nodes.

When multiple domains exist in the network, it is possible to have SR controllers talking to each of the domains and a higher level controller/orchestrator managing the

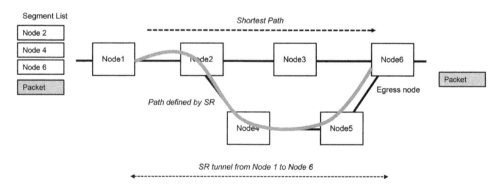

Figure 8.15 Segment routing principle.

end-to-end path. The benefit here is that the end-to-end path can be created through different domains and this can be easier using SR than with inter-domain MPLS-TE. BGP can be used to carry link-state information from the IGPs to a central PCE.

Segment routing lends itself to many use cases as a light way of traffic engineering as now a path through the network is enforced. For 5G network slicing a slice-specific path may be needed e.g. to guarantee a maximum delay or simply to select a path using certain nodes in order to isolate traffic. A tunnel can also be created for other purposes.

With service function chaining, certain sets of nodes or containerized functions in cloud deployments may be selected that perform specific tasks, like firewall or deep packet inspection as examples, after which the flow is delivered to the next function in the service chain.

8.2.5 IP and Optical

The underlying physical medium for IP links is supplied by optics (fibres and wavelengths) or wireless links; wireless mostly in the last-mile access links. A physical media layer complemented with IP brings a combination of high bandwidth, high granularity direction of flows and also, in general, services from the IP domain. Basically, traffic can be routed on the optical layer, on the IP layer or on both.

In Figure 8.16, gNBs attach to the network via an IP router over an optical layer, where there exists some redundancy with optical switches that are capable of switching at λ or fibre level. This level of switching is coarse-grained, so connectivity is not flexible at the optical layer alone. Many approaches exist. IP topology usually does not match optical topology, depending on requirements for the optical layer and possibilities for the routers to provide directly the optical link.

G-MPLS as mentioned covers the optical layer, defining optical LSPs, and this extends a control plane to the optical layer, which then allows a common control plane. An alternative for G-MPLS-based cross-layer coordination is as in Figure 8.16, with the help of SDN (Software Defined Networking) type of control and/or a management system that interfaces both domains. For TE services, information is collected from the network and then a path (or multiple paths) created through it. The central controller can collect information

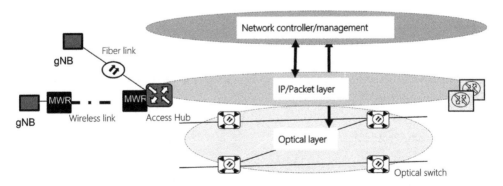

Figure 8.16 Connectivity with IP and optics.

for both the optical layer and the packet layer(s). In Ethernet-based fronthaul, the optical layer could be combined with Ethernet switches according to the fronthaul profile/TSN, basically with similar considerations as with IP and optical.

8.2.6 IPv4 and IPv6

For the IP layer itself, two protocols exist: IPv6 and IPv4 [12, 13]. Gradually, more and more networks use IPv6 and one option is to build 5G transport with IPv6 instead of the IPv4 protocol. The 3GPP protocol stack definition for 5G logical interfaces includes both IPv4 and IPv6, with preference for IPv6. O-RAN also includes both IPv4 and IPv6 in the fronthaul definitions, in case of the IP stack. IP VPNs support both IPv4 and IPv6 (VPNv4 and VPNv6).

IPv6 has similarities with IPv4 (e.g. longest match prefix routing), but it also includes major improvements and new designs. Large addressing space is a well-known topic: the IPv6 header includes 128 bits for an address, with 64 bits for the network portion and another 64 bits for the host portion (interface ID). Other improvements include avoidance of broadcasts (which are replaced with multicast), use of link local addresses, retiring of ARP, removal of checksum from the IPv6 header, and so on. Co-existence with IPv4 is supported with a dual stack.

Nodes in mobile backhaul (and fronthaul) networks do not generally need to be accessible from the public internet (apart from the UPF interfacing the public internet). Still, the huge address space of IPv6 is a benefit. Large mobile backhaul networks have a high number of nodes, including 5G FR2 small cells, and for fronthaul there can be huge numbers of remote radios (O-RUs) that need IP connectivity – for management purposes or for the O-RAN-defined IP transport option.

With IPv4 it is often the case that the previously designed IPv4 address planning is not optimal any more with 5G, summarization does not work, or there are regions of the network with overlapping IPv4 address spaces. With multiple radio technologies on the same site, and possibly a transport network that is shared with many operators, there is a need for many IPv4 addresses on each site. This is one example where IPv6 is of benefit.

Routing protocols include support for IPv6 prefixes and also support for distributing both IPv4 and IPv6 prefixes. With IPv6, routing protocols like OSPF can operate on IPv6 link local addresses, which is different from the IPv4 behaviour. Also configured addresses may be used, to have better visibility to the network nodes.

IPv6 technology also solves issues on the UE side, although the UE IP layer is not visible to the transport network as the user IP is tunnelled with the GTP-U protocol. UEs like IOT terminals need connectivity to the servers in the internet only once in a while (e.g. for reporting a measurement or condition) and benefit from IPv6 since a permanent, globally routable IPv6 address can be allocated.

8.2.7 Routing Protocols

Routing protocols are essential for advertising local networks and discovering reachability to remote networks, and for supporting recovery from failures in the network links and nodes, as well as within a data centre, for connectivity between the containers/servers and data centre switches. Routing protocols are also needed for network discovery and in this way they help automate network operations.

Routing protocols each have their own merits and which protocol is selected depends on the needs of the network but also in practice of the existing network and existing competencies, as rolling out a new protocol requires learning. Depending on the network, the RAN side has in many cases had less exposure to IP than the core network. This is changing with 5G, as it gets increasingly difficult to manage the 5G access network and meet all the requirements for backhaul and fronthaul without IP services for routing and recovery.

A new area within backhaul and fronthaul is cloud deployment and related data centre designs which have connectivity needs both within the data centre and externally to other sites. A routing protocol can recover connectivity in case of failure of a data centre switch or link.

The most relevant of the protocols from 5G transport aspects are perhaps BGP, OSPF and IS-IS. Interior gateway protocols like OSPF and IS-IS are link-state protocols. BGP is for inter-AS (Autonomous System) routing and used for the the public internet. MP-BGP is a key building block for services like IP VPNs and Ethernet VPNs. BGP is also extended to carry link-state information from the IGPs to a central controller, which is a way of supporting network programmability. BGP is also used in data centres.

OSPF and IS-IS maintain local knowledge of topology and link states of a network area. The topology database is kept up-to-date and an active topology is calculated using Dijkstra's shortest path first algorithm. Extensions have been created (for traffic engineering, segment routing, etc.).

In a host (e.g. gNB), a routing protocol can be used to advertise local networks to other routers and in this way allow automated operation. Due to 5G network slicing and different forms of RAN and transport sharing, there may be multiple local subnets. If there is no routing protocol, the routes are statically configured, which is an operational effort. The main alternative is an effective SDN control or management system.

For backhaul, it is common to group base stations from an area into an access hub site using Ethernet point-to-point links or Ethernet services, and have the hub connected to an IP VPN service. This hub device acts as a CE router which attaches the customers (networks of the gNBs) to the IP VPN via the PE node. Preferably some routing protocol runs between the CE and PE, which could be one of those mentioned – BGP, OSPF or IS-IS.

In the TSN task group of IEEE, IS-IS is included in some of the enhancements for topology discovery. For the fronthaul profiles however, IS-IS is not required. IS-IS was introduced in the LAN/MAN area to solve shortest-path bridging-type data centre use cases.

Routing protocols detect failures and find alternative ways to reach the destination. Gradually, in mobile networks, expectations from users have grown for even the basic mobile broadband service to a level where the service is expected to be continuously available. Outages are poorly tolerated and without automatic recovery from a link or node failure, repair times are long. An example calculated in Chapter 4 shows the dramatic improvement to availability when manual repair is replaced with automatic recovery. Pre-requisites are that the network has multiple alternative paths and the use of a routing protocol (or some other protocol) to recover automatically. Physical links are needed more and more with 5G not only due to resiliency but also due to small cells, and there are new alternatives to support those with wireless transmission (e.g. E-band radios) and IAB (Integrated Access and Backhaul). Having more connectivity options increases the benefits of routed networks.

Detecting failures can be almost immediate in case of a local physical signal down event, which already covers cases like (local) cable cuts or power outages on the remote end device, causing loss of signal at the local end. In general, detection requires more time and relies on some protocol, which could be a routing protocol hello or some other keepalive message. BFD (Bidirectional Forwarding Detection) is commonly used for faster failure detection. BFD can also be used to withdraw a static route, which as a special case allows simple dynamism in the network without running a routing protocol.

In case of a link-state protocol, the state information – such as the occurrence of a link down event – is propagated thorughout the area and new shortest paths are calculated and installed to the forwarding table. All of these actions take some time. Depending on the size of an area and the number of nodes, it could be sub-second to some seconds. The actions required are summarized in Table 8.1

Routing protocols have been and are continuously being enhanced in several areas to support new services and use cases. Some of these enhancements are as follows:

- Traffic engineering extensions help set up a path through the network, which is useful for cases where a specific SLA (Service Level Agreement) and path are needed.
- Multiprotocol BGP enables advertising IPv4 and IPv6 and also Ethernet MAC addresses, which are needed for the E-VPN Ethernet services.
- Segment routing extensions.
- IS-IS has been included in the IEEE TSN standard to help topology management and covers MAC addresses also.
- BGP multipath capability.
- BGP is enhanced to carry link-state information from the IGPs and can interface SDN controller or other central element that needs topology and other information from the nodes.

8.2.8 Loop-Free Alternates

When a failure on a link or node is detected, IGPs propagate the failure information to other routers in the area, after which a new shortest path topology is calculated by each router and installed into the active forwarding table in each router. This takes some time and meanwhile, packets via the failing link or node are dropped.

A loop-free alternate (see RFC 5714 [15] for framework definition) can shorten the above process, avoiding or minimizing the duration of blackholed traffic by having a

Table 8.1 Recovery with routing (RFC 3469) [14].

	Start	End
Fault detection time	Network impaiment	Fault detected
Fault hold-off time	Fault detected	Start of notification
Fault notification time	Start of notification	Start of recovery
Recovery operation time	Start of recovery	Recovery completed
Traffic recovery time	Recovery completed	Traffic recovered

pre-calculated back-up path that is used in case of a local failure being detected. When the current next hop fails it is removed from the active forwarding table and the selected loop-free alternate is taken as the new next hop, so packet flow via the failing link now reaches the destination via the new next hop. In parallel, recovery process in the area continues and a new shortest path first topology is calculated which possibly leads to changes in the active forwarding topology, but meanwhile traffic flow can use the Loop-Free Alternate (LFA) next hop. An example is shown in Figure 8.17.

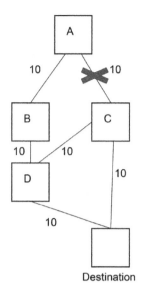

Figure 8.17 Loop-free alternate example.

In the simple example of Figure 8.17, all link costs are 10. Before failure, the shortest path from A to the destination network is via router C, with the sum of costs on the path A–C– destination = 20. If the link A–C fails, router B can serve as an alternative next hop. Router B has its shortest path to the destination via D and this clearly avoids the failed link A–C. Also, a loop topology is not formed.

In more complex topologies it is not always possible to find alternative loop-free next hops from adjacent neighbours. Remote nodes can also act as LFA next hops, as in RFC 7490 [16]. In that case a tunnel is created between the local node and the remote LFA. The tunnel could be created with MPLS, or GRE for example. In case of failure, the remote node via the tunnel acts as the LFA next hop.

Main benefit is that the duration of the break is shortened. The limitation is as mentioned – not all failures and topologies can be covered. As the decision is local, there is no need to wait for the failure propagation and SPF (Shortest Path First) calculation but instead the alternative next hop can be used directly after the failure is detected. Compared to MPLS FRR (Fast Re-Route), the approach is simpler yet with a similar level of performance (sub-50 ms), which increases the availability of transport connectivity in 5G networks. If a link-state protocol is in operation, introducing LFAs may be a simple way to achieve the performance of MPLS FRR.

8.2.9 Carrier Ethernet Services

Ethernet services are an important part of 5G backhaul and fronthaul connectivity, in the form of point-to-point Ethernet links using fibre or wireless media, or as other Ethernet services. Metro Ethernet (MEF) has definitions where services are abstracted, so that they become implementation-agnostic. What was previously discussed with E-VPN does not mean that E-VPN would be mandated for Carrier Ethernet services , it is merely an implementation option.

Depending on the case, MEF services could be implemented on a point-to-point dedicated fibre or wavelength, or it could be a more complex network, such as with IP/MPLS technologies, over pseudowires for point-to-point service or VPLS for multipoint service, or more recently with E-VPN, due to the benefits discussed – especially in resilient attachment to the service. Basic service types are as in MEF 6.1 Ethernet Services Definitions – Phase 2 [17]. An example application is shown in Figure 8.18 for connecting gNBs to a hub node.

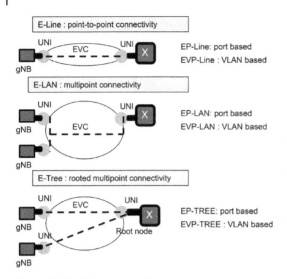

Figure 8.18 Ethernet services.

EP-Line (Ethernet Private Line) emulates point-to-point Ethernet. EP-LAN (Ethernet Private LAN) emulates multipoint LAN. EP-Tree (Ethernet Private Tree) emulates rooted tree, where one node is the root and has connectivity to the leaves (and leaves have connectivity to the root), but the leaves cannot talk to each other. Most commonly, E-Line and E-LAN have been deployed. Further separation comes with EP-Line and and EVP-Line (Ethernet Virtual Private Line), where the former means VLAN-unaware service and the latter means VLAN-aware service. The same distinction goes for E-LAN and E-Tree.

MEF has extended the definition to cover 5G fronthaul, meaning the 5G high-layer split point (midhaul in this text). The issue with low-layer fronthaul is the strict latency requirement, which is difficult to support unless that service is given a dedicated resource (fibre or wavelength), or latency is guaranteed with the help of TE methods including TSN.

Ethernet connectivity is not mandated for any logical interface, yet MEF services offer a benefit when the customer needs a lower-level (below IP) connectivity service that avoids IP routing peers with the service provider. Fronthaul Ethernet connectivity is needed in the Ethernet variant of the eCPRI. There, as mentioned, the difficulty in deploying the MEF service may be the availability of suitable SLAs, especially for a strict enough latency class.

8.2.10 Ethernet Link Aggregation

With carrier-grade resiliency requirements for 5G backhaul and fronthaul combined with the need to have efficient use of links in an active–active manner, Ethernet link aggregation is useful, as it allows active–active forwarding for use cases like load sharing and resiliency. Link aggregation can also provide more capacity as it combines multiple links into an aggregate. It is defined in IEEE 802.1AX [18].

The link aggregation group consists of individual link group members, to which a distributing function distributes flows using a hash algorithm. Commonly, only conversation/flow-based distribution is used, as packet-by-packet distribution may create issues with packet disorder. Flow-based distribution ensures that packets belonging to one flow are kept on the same link.

Distribution requires some entropy in the header field, which may not always be trivial with tunnelled traffic, as is the case with GTP-U or IPsec, since the original IP packet is not directly visible. Commonly, hashing algorithms use IP source and destination addresses,

MAC source and destination addresses, source and destination ports and VLAN IDs, and in many cases these provide enough entropy to be useful.

One big benefit of link aggregation is that all links can simultaneously be used actively, which is not supported by native Ethernet bridging. In bridging, only one link is used in the active topology that is enforced by the IEEE spanning tree protocol, which is a distance-vector protocol. With link aggregation, both links can be active, allowing active–active dual homing to two PEs (edge devices). This makes link aggregation a useful feature in attaching Ethernet access to Ethernet services.

Link aggregation includes a control protocol, LACP (Link Aggregation Control Protocol). In addition to the active–active mode, it is possible to operate in active–passive mode.

8.3 Securing the Network

8.3.1 IPsec and IKEv2

The IP security suite of protocols (IPsec, RFC 4301 [19]) is the 3GPP selected technology on network layer for cryptographic protection. IPsec services include data origin authentication, integrity protection, encryption and anti-replay protection. Network domain security (NDS/IP) in specification 3GPP 33.210 [20] was originally developed with a focus on control plane protection. Security architecture (33.501 [21]) refers then to logical interface protection (also for the user plane) to the NDS/IP and in practice definitions are used for user plane transport interfaces as well. The logical interface protocol stacks are often shown without the network domain protection (without IPsec), but ensuring security is an essential part of the transport network layer even if not visible in all of the interface definitions. A summary of the cryptographic protection available in a 5G system without NDS/IP on different logical interfaces was presented in Chapter 4.

The specification in 3GPP for NDS/IP is not intended to cover all security aspects that are required in protecting mobile backhaul and fronthaul. Instead, it focuses on defining cryptographic protection tools with IPsec (and TLS) to complement the protection built into the 5G radio and core, NAS signalling and air interface. For transport links, network elements and nodes attached, additional methods have to be deployed to keep the network available and secure in case of accidental misconfigurations or malicious attempts. These topics include:

- Disabling any transport or connectivity service, node or port that is not needed
- Disabling and blocking flows that are not needed
- Restricting access to all network resources unless explicitly allowed and authorized
- Protecting control and user plane computing resources of nodes by ingress policing
- Detecting and recording anomalies and possibly acting on those instantenously

NDS defines a security domain, which is a network of one operator or a network that is administered by a single entity. Between security domains, IPsec tunnel mode is required and within a security domain transport mode is also allowed.

IKEv2 (Internet Key Exchange) (RFC 7296 [22]) is the control plane protocol for IPsec. It manages key exchange and sets up Security Associations (SAs). IPsec SAs are defined with a Security Parameter Index (SPI), an IP destination address and a security protocol identifier. A Security Policy Database (SPD) manages the policies in terms of which flows shall be protected and by which means. A single IPsec SA may carry multiple IP flows (e.g. 5G control plane flow – NG-AP) and user plane (N3) flows, or then separate IP sec tunnels may be created. At the extreme, all flows from a single gNB may use the same tunnel, or then a tunnel may be created for the different traffic types. This all depends on the implementation.

In the single-tunnel case, if the flows have different destinations, a security gateway is needed to terminate the tunnel before the individual flows can be routed to their destinations. For a remote destination over a non-trusted network, a new IPsec tunnel is needed. An alternative is to set up a separate tunnel per destination from the source.

A profile is defined in 3GPP 33.210 for both IPsec and IKEv2, in order to help interworking across domains. The profile refers to the ESP protocol (RFC 4303 [23]) with extended sequence numbering. Since, over security domains, security gateways (SEGs) are used, tunnel mode is mandatory. Authentication and encryption transforms are defined in RFC 8221 [24]. For IKEv2 similarly, a profile is given.

With IPsec gateways (SEGs), traffic flows are protected with IPsec tunnel mode, meaning that the whole original IP packet becomes an inner IP packet that is encapsulated in another IP, IPsec packet. The outer IP tunnel endpoints are then at the SEGs, or they may also reside at the hosts, such as gNB. The transport mode is from host to host and leaves the original IP header visible.

5G introduced a disaggregation alternative for the RAN and related to that, network functions are deployed in data centres in a flexible way. In this case, IPsec is either integrated with the network function or then a central IPsec tunnel endpoint (or endpoints) are used.

Cryptography evolves, and so does the computing capability – with new possibilities to try to break the system. Algorithms used have to be kept up-to-date with the latest standard and developments in computing have to be followed up. Quantum computing is a new technology which opens new ways for solving mathematical problems that were previously considered technically unfeasible to be solved in reasonable computing time. Research is ongoing in the area of quantum-resistant ciphers.

3GPP refers to IETF RFC 8221 for cryptographic algorithms. AES-GCM with 16 octet ICV shall be supported, which supports authenticated encryption.

A central SEG terminates IPsec tunnels from many base stations, and any failure of the SEG (or any key link to the SEG) leads to outage of that area. With a back-up SEG architecture, another SEG takes over the role of the active SEG. As IPsec tunnels maintain the state, failover to another SEG requires either synchronizing that state actively between the primary and back-up SEG, or the switchover is a hard one and IPsec tunnels need to be re-etablished.

For geo-redundancy, this back-up SEG is located on another site. This creates complexity for maintaining the state and in this case a hard switchover is usually needed, which then means losing the existing PDU sessions during the failure.

8.3.2 Link-Layer Security (MACSEC)

The link-layer security solution is specified by IEEE (MACSEC) with a related IEEE 802.1ax Port Based Authentication and Security Device Identity (IEEE 802.1ar) [25–27]. MACSEC-based solutions complement the IPsec solutions in 5G transport as a technology that addresses link-level security concerns, while IPsec addresses end-to-end at networking level.

Figure 8.19 illustrates a case where IPsec tunnels protect traffic end to end between the gNBs and core network peers. MACSEC augments protection at the link layer between multilayer switches as in the example. In Figure 8.19, MACSEC could also extend in a hop-by-hop manner into the core of the network if needed.

For high-capacity cryptographic services, the benefit of MACSEC is that it is typically implemented directly at the data link layer close to the physical interface in hardware, while IPsec is commonly implemented centrally in a node. In this way, MACSEC capacity scales more easily and can achieve higher performance than IPsec. Also adding ports is simpler, since as the MACSEC engine is port specific, capacity increases with port count. For IPsec, adding high-capacity flows via new ports to a node would require adding capacity to the central IPsec engine. With increasing port rates and port counts, IPsec starts to have limitations concerning capacity. Low-layer fronthaul is an example where both port counts and throughputs are especially high, but IPsec throughput bottlenecks may also exist in 5G backhaul or midhaul.

Basically, MACSEC secures hop by hop. This makes the full original packet available in clear text for processing at each MACSEC-enabled node (e.g. a router). Also, IP control planes like routing protocols are protected over the link but are available at the node as clear text for processing, as well as other protocols.

The hop-by-hop nature of MACSEC is a drawback in another sense, as each leg will have its own MACSEC protection and successive encrypt/decrypt operations take place – introducing latency in a scenario of multiple hops.

MACSEC was originally specified by the IEEE for the LAN/MAN (Local Area/Metropolitan Area Network) environment. However, with Ethernet services, one Ethernet

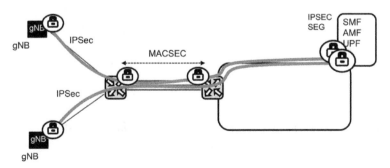

Figure 8.19 MACSEC and IPsec.

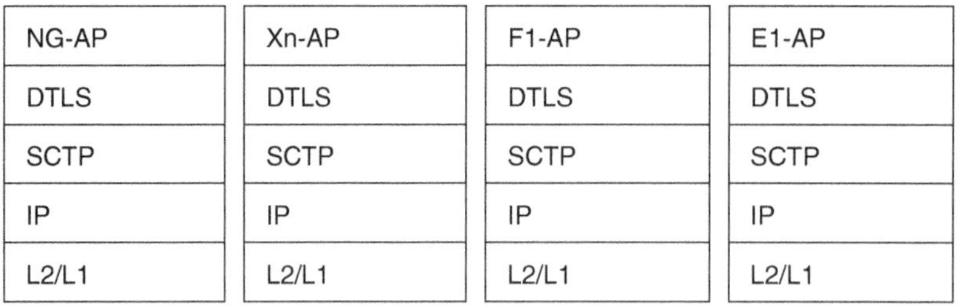

NG-AP	Xn-AP	F1-AP	E1-AP
DTLS	DTLS	DTLS	DTLS
SCTP	SCTP	SCTP	SCTP
IP	IP	IP	IP
L2/L1	L2/L1	L2/L1	L2/L1

Figure 8.20 DTLS-based protocol stacks [21].

segment may encompass multiple devices and links, even if it appears as a single segment for the customer.

Additional enhancements include:

- Having an 802.1Q VLAN ID field in the clear in order to support VLAN-specific services
- An extended sequence number to avoid frequent re-keying
- A re-order buffer in order to allow occasional out-of-order frames

In 5G fronthaul (eCPRI/O-RAN) bit rates are high and, due to this, MACSEC has a benefit when compared to IPsec due to the port-based crypto capability. As the PDCP protocol layer extends between the gNB and the UE, user communications (in the form of the IQ stream) are protected by the PDCP even without transport-layer protection. O-RAN fronthaul specification does not define specific cryptographic protection.

8.3.3 DTLS

With 5G, DTLS (Datagram Transport Layer Security) has been added to the SCTP-based control plane interfaces (N2, Xn, F1, E1) for additional protection, so it is not precluding the use of IPsec. DTLS over SCTP is specified in IETF RFC 6083 [28], based on TLS and used for mutual authentication, integrity protection, replay protection and confidentiality protection. DTLS 1.2 is mandated, DTLS 1.0 shall not be supported.

DTLS-based protocol stacks are shown in Figure 8.20 for N2, Xn, F1 and E1.

DTLS has a benefit in having a better fit with the client–server architecture model than the use of IPsec tunnels, which require establishment of a secure association (a tunnel) with the help of IKEv2 and maintaining the tunnel. While a secure association needs to be set up with DTLS too, DTLS operates closer to the application and as such can more easily respond to requirements from individual applications – especially in the case where the applications are located (and re-located) dynamically in the cloud.

8.4 Time-Sensitive Networking and Deterministic Networks

8.4.1 Motivation for TSN

Both Time-Sensitive Networking (TSN) and Deterministic Networks (DetNet) aim at deterministic characteristics for packet delivery – especially in environments where there are critical time-sensitive flows mixed with other flows, TSN for Ethernet and DetNet for IP/MPLS.

Today, for example, many industrial automation-type use cases rely on a dedicated network and specific technologies such as ProfiNet or Ethernet/IP. TSN targets meeting these and other requirements that focus on guaranteed latency and reliability of the service.

If standard Ethernet (with TSN enhancements) can be used for these applications, it allows an evolution into a common technology base for industrial use cases. A further potential benefit lies in combining the TSN and other services into a single network so that TSN becomes a common network technology in segments needing TSN-type of services. This requires that other traffic flows do not compromise the performance of TSN flows (streams in TSN terms).

TSN itself has a special role related to 5G, as one operating mode has a 5G system as a bridge that supports TSN services. 5G as a bridge is a 5G system-level topic, nevertheless the technology is from the transport network layer. The other application in 5G is for fronthaul, for which purpose a specific fronthaul profile (IEEE 802.1CM [29]) is defined and, in general, URLLC service may require TSN services on transport links. TSN is driven by IEEE and DetNet by IETF, however, in close cooperation.

In packet-switched networks there is generally no guaranteed upper bound to delay. Delay experienced by packets follows some distribution, which in the case of congestion makes part of the packets get delayed excessively ('long tail'). If the application essentially needs all packets arriving in a timely manner, basic packet switching is in many cases inadequate. For reliability of the service, there should also ideally be no packet loss due to other reasons, like failures on links and nodes or bit errors.

The TSN group (see Ref. [30]) has defined multiple enhancements to IEEE standards to ensure that requirements for deterministic service can be met. These can be grouped into four areas:

- synchronization for the purpose of establishing a common time base
- scheduling enhancements
- packet replication for reliability
- management of the TSN network

The enhancements form a tool box where not all components are needed for every application, but rather features are deployed as needed to ensure proper characteristics for the service. Additionally, application specific profiles have been created, like the fronthaul profile (IEE802.1CM). TSN originally started as the Audio and Video Bridging (AVB) group, but since that many other applications have been considered and the group was renamed accordingly. Only few of the TSN technologies are discussed here; namely, the fronthaul profile, frame pre-emption, frame replication and elimination and TSN management. DetNet is covered in the last section.

8.4.2 IEEE 802.1CM – TSN for Fronthaul

An obvious application for 5G and TSN is fronthaul, and a separate IEEE standard exists (IEEE 802.1CM). This standard includes two profiles, where profile A does not use any advanced Ethernet TSN features. Profile B uses frame pre-emption. Profile A is mandatory to be supported, Profile B is optional. Additionally, 802.1CM includes requirements for synchronization, which are covered in the synchronization sections.

IEEE 802.1CM Profiles A and B
Related to queueing and scheduling:

- Profile A. Essentially mapping critical flows (fronthaul) with high-priority Priority Code Point (PCP) and in this way ensuring these flows are receiving the service they need with no interference from other flows.
- Profile B (optional). Adding the use of frame pre-emption (from the TSN family of standards).

Profile A uses existing Ethernet capabilities on queueing and scheduling, and this is the mandatory part of the fronthaul profile. Strict priority queueing is required.

High-priority flow is given priority, so that delay contribution due to congestion with higher-priority frames does not occur. What can still happen is that multiple of the highest-priority frames arrive at ingress ports and compete for transmission out of the same egress port, as in Figure 8.21 for the traffic flows sourced by the two O-RUs. For fronthaul, almost all traffic is high priority, with M-plane as the main exception. This has to be taken into account in planning, so that it is known at the design phase what is the worst-case situation that may take place, considering arrival of the ingress flows and then the egress service rate.

Ongoing transmission of a low-priority frame out of an egress port also means that the highest-priority frame targeting the same egress port needs to wait until the ongoing frame transmission is completed. Frame size is defined to be 2000 octets maximum for the ports supporting fronthaul traffic.

In order to guarantee latency, network topology, physical distances in the links, number of hops and nodes, the above factors are factored in to confirm that the worst-case delay value is still within the latency budget. The optional part (included in Profile B but not in Profile A) is frame pre-emption. This is especially useful if there are large non-critical frames in the traffic mix, multiple bridges on the path and the link speeds are low.

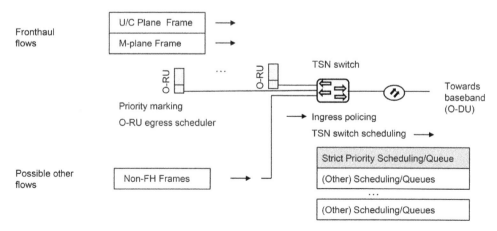

Figure 8.21 Profile A with strict priority scheduling.

8.4.3 Frame Pre-emption

Frame pre-emption is defined in IEEE 802.3br Interspersing Express Traffic and 802.1bu Frame Preemption [31, 32]. With frame pre-emption, the ongoing transmission of a pre-emptable frame can be stopped, the express frame transmitted and then transmission of the pre-emptable frame (frame fragment) resumed.

In Figure 8.22, the upper part of the diagram shows the case without pre-emption. A fronthaul high-priority frame is transmitted only after the ongoing transmission of the background frame is completed. For a 2000 octet frame, the transmission time on a 10 Gbps link is 1.6 μs, so there is some optimization possible. An additional benefit is in reducing latency variation for the critical frames.

In the lower part of the diagram the TSN switch stops the ongoing transmission of the background frame, transmits the high-priority frame and then resumes transmission of the background frame. Some overhead is caused, as the background frame is now transmitted in two pieces and also an inter-frame gap is required. It is also possible in advance to hold the start of a pre-emptable frame transmission, in favour of an incoming express frame (a hold-and-release procedure).

Frame pre-emption operates using a division into Express MAC and Pre-emptable MAC sublayers with the related MAC merge sublayer (Figure 8.23).

The next TSN bridge receives the fragments and reassembles the original frame. Frame pre-emption capability needs to be supported by the peer.

8.4.4 Frame Replication and Elimination

Frame Replication and Elimination (FRER) was mentioned in Chapter 4 as one technology that can increase service reliability over transport links, especially considering URLLC services that require ultra-reliability. As such, FRER is potentially relevant for both backhaul and fronthaul. It is not part of the IEEE 802.1CM fronthaul profile and is defined as IEEE 802.1CB [33].

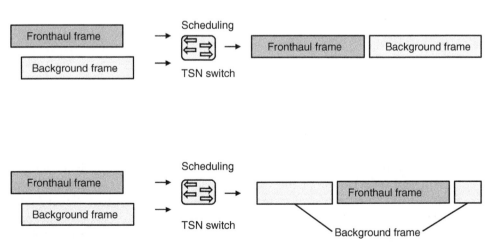

Figure 8.22 Frame pre-emption.

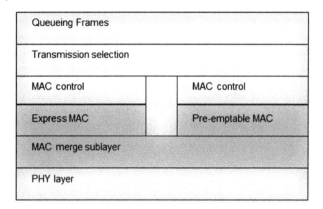

Figure 8.23 MAC merge sublayer [32].

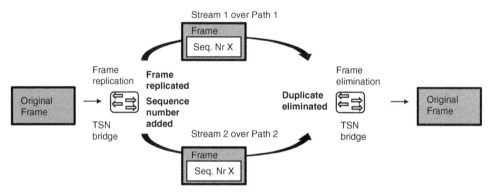

Figure 8.24 FRER application.

FRER requires ideally disjoint and independent transport paths over which frames are replicated so that the same frame is sent simultaneously over the paths as in Figure 8.24. Failure on one path does not impact the service as long as the same frame arrives via the other path. There is no loss of even a single frame, which would normally happen in case of a recovery mechanism. The duplicate frame or frames are eliminated with the help of a sequence number.

The FRER function can reside in an end system or in a relay system (bridge), enabling a number of configurations – either by replicating between two end systems, on some segments in the network between two bridges, or some combination of these. It is also flexible in that only some of the streams (critical ones) are replicated but others are not. This saves bandwidth on the links compared to the case of doubling the capacity for all traffic flows and is especially useful if the critical traffic in the TSN network is of low volume compared to the background traffic which would not need to be replicated. For many TSN use cases this is true.

8.4.5 Management

TSN data flow is called a 'TSN stream', which is identified by fields like IP and MAC source and destination addresses and VLAN IDs. The TSN network has to know the needs that each stream has and then reserve a path through the TSN network with resources allocated for that stream. A talker node is the source node and the stream can be consumed by one or multiple listener nodes. Three different models are provided in 802.1Qcc [34]: fully distributed model, centralized network/distributed user management and fully centralized model.

In the fully distributed model, end stations use a protocol like the Stream Reservation Protocol (SRP) for signalling over the UNI (see Figure 8.25).

User/network configuration information flows through each of the nodes. The TSN nodes reserve resources by local decisions and there is no network-level entity for configuration.

In a distributed user/centralized network model (see Figure 8.26), end stations use the protocol over the UNI as in the fully distributed case but configuration of the TSN nodes (bridges) is done centrally, using a management interface.

The centralized network management function (that could be implemented on one of the bridges or end stations or be a standalone element) has knowledge of the whole network and all streams and can optimize the network accordingly. It can also perform more complex computations. The management interface between the TSN nodes and the centralized network configuration node could be Netconf/Restconf or SNMP (Simple Network Management Protocol), for example.

In the third alternative the user management is also centralized, leading to a fully centralized user/network configuration model (Figure 8.27).

In the fully centralized model, both user and network configurations occur centrally, and there is no need for a signalling protocol over the UNI. This model has the additional benefit that now there is also a central entity for the user configuration, as the user nodes (talkers and listeners) may also require a lot of configuration and coordination (e.g. for setting up the streams and arranging the transmission times if needed). The configuration

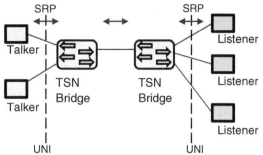

Figure 8.25 Distributed model example [34].

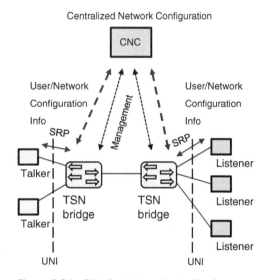

Figure 8.26 Distributed user/centralized network configuration example [34].

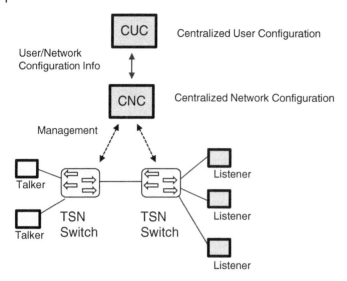

Figure 8.27 Fully centralized model example [34].

protocol and info model for the user node configuration are left out of the scope, which allows vendor-specific user node configurations. For the TSN bridges, the YANG data model is defined. The TSN UNI is now between the two central entities, CUC and CNC.

The fully centralized model is in practice the applicable one for fronthaul, since there is no signalling protocol defined to be used but instead, the user nodes (O-RUs and baseband) and the network nodes (TSN bridge) are configured via management, this of course being an implementation topic, too.

8.4.6 Deterministic Networks

The DetNet working group in IETF [35] is closely related to the IEEE task group TSN and also developed in coordination. DetNet focuses on IP and MPLS networks that are under single administration, while TSN's scope is the LAN/MAN as for the IEEE. An example case for DetNet might be a smart grid.

As with TSN, a lot of work has been done and is ongoing in IETF, with work started later than in IEEE. In 3GPP backhaul and fronthaul, logical interface definitions for 5G backhaul or fronthaul as such do not mandate or expect DetNet services. URLLC use cases may require DetNet-type service from the transport network. The DetNet data plane is divided into two sublayers – a services sublayer and a forwarding sublayer. The services sublayer mainly provides for packet replication, elimination and ordering.

Many approaches are possible with DetNet in the data plane, and in control and management. Some of the services rely on a separate lower-layer subnetwork, like TSN or MPLS TE. Data plane alternatives include IP (no services sublayer defined, RFC 8939), MPLS (RFC 8964) and IP over TSN (RFC 9023) [36–38]. Service protection, in the form of replication and elimination, is supported by TSN and MPLS.

The forwarding sublayer supports resource reservations, and explicit paths and technologies include IP routing, MPLS and Ethernet bridging. The lower layer, like Ethernet, needs to be aware of the flow (flow mapping). For scheduling, shaping and queuing, it is assumed that many of the services developed for bridging in the TSN group can be supported by routers as well. Required reservations are created with the help of controllers or by control plane protocols in a distributed way. Flow characteristics need to be available so that the required resources are known and can be allocated.

8.5 Programmable Network and Operability

8.5.1 Software-Defined Networking Initially

Networking equipment like switches and routers generally have data plane (fast packet forwarding engine), control plane (control protocols processed in slow path by CPU) and management plane (also handled in slow path). One of the original targets of SDN has been to separate the data plane from the control plane and program the data path flow via a separate central controller. This SDN controller hosts the intelligence while the data plane networking device is simple as it only gets to program the forwarding plane with instructions that are received from the controller. The instruction is, for example, a forwarding table entry telling the device to egress a frame with destination address X out of port Y, so it is rather low-level information. Programming all these forwarding entries from the controller means that there is no need for the device to run a control plane protocol (like a routing protocol) or learn by flooding (like with native Ethernet). In essence, it is changing the architecture from one where the control plane is distributed to the elements to one where the control plane is central. At the same time, it changes the network element architecture as now that device no longer appears as a monolithic box handling both control plane (and related intelligence) and data plane. Furthermore, if the data plane operations and the interface to the controller are well standardized, the same controller can talk to any box that understands this interface and supports the instructions. An example is Open Flow [39].

When network flows can be programmed centrally by the controller, provisioning of a new service flow is fast, as instead of configuration commands on each of the elements in the chain separately (possibly with different management systems), a whole end-to-end path can be set up from the same controller, of course assuming all devices in the path support this. For the user, this means a shortened waiting time for the connectivity service.

For SW developers requiring networking functionality, immediate instantiation of connectivity service is of value, as SW can then be developed and tested continuously (a topic that is relevant for cloud SW development). However, often an existing network is disparate in terms of old and new equipment, typically from multiple vendors, and the devices seldom match the ideal SDN architecture – with strict separation of data and control planes. Furthermore, in a wide area network, many existing designs successfully support resiliency and other key features that rely on distributed control planes, so there is inertia and cost involved for changing this. Wide area networks seldom have the opportunity for greenfield deployment, while for data centres this may be the case. Specific functions like fast recovery would be difficult to implement without some intelligence and local decisions in the nodes, since if any change in the forwarding relies on a central controller, there is necessarily some delay involved – first the indication of the failure has to be propagated to the controller, and then the controller has to analyse the situation and respond.

8.5.2 Benefits with Central Controller

Increasing network programmability with an SDN-type approach with a central controller, but without necessarily fully following the original principle, can however lead to big benefits. Link-state routing protocols have been enhanced to carry more and more information than simply the link states, but there are still limits to what information can reasonably be added to routing protocols and then decided locally based on that information.

One example is reacting to load situations. It may be useful to direct some traffic off a congested link or node to another path. Collecting this is simpler to do centrally, and then optimize traffic flows accordingly, taking into account the situation in the whole network and the optimization driven by intelligence in the central element. This, however, requires that real-time (or at least near real-time) network performance data is available to make this decision.

Also, there are likely to be multiple domains – based on technologies like optical, microwave, IP (routing and switching) and further based on administration. For IP and optical, a common G-MPLS control plane was developed for traffic engineering. Another way to address the same issue is to have a separate controller per domain and on top of that a cross-domain coordination, as in Figure 8.28. For topology discovery, BGP can be used [40] to carry link-state and TE information from the IGPs and then interface a central controller or PCE. As an example case, a central controller is useful in 5G network slicing.

Network slicing is a new 5G feature which generally is provisioned over multiple domains – RAN, core and transport. The transport service may consist of multiple legs with multiple

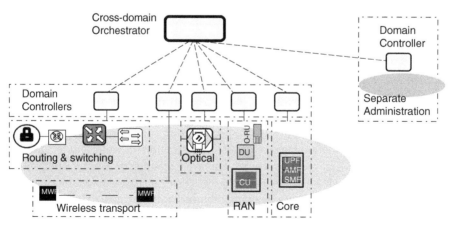

Figure 8.28 Cross-domain operation.

technologies. Different transport legs or layers may be supported by dedicated domain-specific controllers. Without a central entity (network orchestrator), provisioning has to be done separately for each of the domains. The central orchestrator, on the other hand, abstracts the network and interfaces with a domain-specific controller for carrying out the task.

8.5.3 Netconf/YANG

For configuration of network nodes, Netconf is the protocol and YANG (Yet Another New Generation) provides for the object modelling language and then the actual data models. Netconf/YANG are important since they allow opening of the management of the network. Netconf mandates the use of SSH (or TLS) for cryptographic protection; the protocol stack is Netconf/SSH/TCP/IP/L2/L1. Basic definitions for Netconf and YANG modelling language are in IETF RFCs [41–45]. For the YANG data model, separate RFCs exist.

Operations are based on the client–server model with Remote Procedure Calls (RPCs), where the server is the network node and the client is the management system. XML (Extensible Markup Language) language is used both for the messages and for the data model.

Netconf distinguishes between configuration data and state data. The latter can only be read. Three data stores are defined for configuration data – start-up, running and candidate data stores – out of which the running data store is mandatory to have, containing all configuration data. As the name implies, the start-up data store is used when the device is booting up. The candidate data store holds a potential new configuration, which can be validated before making a change to the running data store.

Closely related to the Netconf protocol is Restconf (RFC 8040 [46]). This is based on http/https and allows web applications to perform management operations defined in YANG. Due to their openness, Netconf/Restconf allow management of network devices which are modelled by YANG, to the extent that devices support the models (switches, routers, optical multiplexers, etc.). The Open RAN alliance (O-RAN) has defined Netconf/YANG-based management for low-layer fronthaul [47]. This extends the openness from pure network devices to O-RAN RUs and related peer (Baseband).

Figure 8.29 illustrates a case with multiple domains managed by Netconf/YANG.

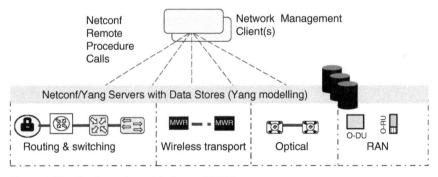

Figure 8.29 Configuration with Netconf/YANG.

References

1 http://www.etsi.org.

2 IETF RFC 4364 BGP/MPLS IP Virtual Private Networks (VPNs).

3 IETF RFC 4760 Multiprotocol Extensions for BGP-4.

4 IETF RFC 7432 BGP MPLS-Based Ethernet VPN.

5 IETF RFC 8660 Segment Routing with the MPLS Data Plane.

6 IETF RFC 8754 IPv6 Segment Routing Header (SRH).

7 IETF RFC 3209 RSVP-TE: Extensions to RSVP for LSP Tunnels.

8 IETF RFC 3473 Generalized Multi-Protocol Label Switching (GMPLS) Signaling Resource ReserVation Protocol-Traffic Engineering (RSVP-TE) Extensions.

9 IETF RFC 8214 Virtual Private Wire Service Support in Ethernet VPN.

10 IETF RFC 8317 Ethernet-Tree (E-Tree) Support in Ethernet VPN (EVPN) and Provider Backbone Bridging EVPN (PBB-EVPN).

11 IEEE 802.1ah Provider Backbone Bridges.

12 IETF RFC 8200 Internet Protocol, Version 6 (IPv6) Specification.

13 IETF RFC 791 Internet Protocol.

14 IETF RFC 3469 Framework for Multi-Protocol Label Switching (MPLS)-based Recovery.

15 IETF RFC 5714 IP Fast Reroute Framework.

16 IETF RFC 7490 Remote Loop-Free Alternate (LFA) Fast Reroute (FRR).

17 MEF 6.1 Metro Ethernet Services Definitions Phase 2.

18 IEEE 802.1AX-2020 – Link Aggregation.

19 IETF RFC 4301 RFC 4301: Security Architecture for the Internet Protocol.

20 3GPP TS 33.210 Network Domain Security (NDS); IP network layer security.

21 3GPP TS 33.501 Security architecture and procedures for 5G System.

22 IETF RFC 7296: Internet Key Exchange Protocol Version 2 (IKEv2).

23 IETF RFC 4303: IP Encapsulating Security Payload (ESP).

24 IETF RFC 8221 Cryptographic Algorithm Implementation Requirements and Usage Guidance for Encapsulating Security Payload (ESP) and Authentication Header (AH).

25 IEEE 802.1AE: MAC Security (MACsec).

26 IEEE 802.1X Port Based Authentication.

27 IEEE 802.1AR: Secure Device Identity.

28 IETF RFC 6083 Datagram Transport Layer Security (DTLS) for Stream Control Transmission Protocol (SCTP).

29 IEEE 802.1CM Time-Sensitive Networking for Fronthaul.

30 https://1.ieee802.org/tsn.

31 IEEE 802.3br Interspersing Express Traffic.

32 IEEE 802.1Qbu – Frame Preemption.

33 IEEE 802.1CB Frame Replication and Elimination for Reliability.

34 IEEE 802.1Qcc Stream Reservation Protocol (SRP) Enhancements and Performance Improvements.

35 https://datatracker.ietf.org/wg/detnet/about.

36 IETF RFC 8939 Deterministic Networking (DetNet) Data Plane: IP.

37 IETF RFC 8964 Deterministic Networking (DetNet) Data Plane: MPLS.

38 IETF RFC 9023 IP over IEEE 802.1 Time-Sensitive Networking (TSN).

39 https://opennetworking.org/about-onf/onf-overview/onf-overview-2014.

40 IETF RFC 7752 North-Bound Distribution of Link-State and Traffic Engineering (TE) Information Using BGP.

41 IETF RFC 6241 Network Configuration Protocol (NETCONF).

42 IETF RFC 7950 The YANG 1.1 Data Modeling Language.

43 IETF RFC 6242 Using the NETCONF Protocol over Secure Shell (SSH).

44 IETF RFC 4252 The Secure Shell (SSH) Authentication Protocol.

45 IETF RFC 4253 The Secure Shell (SSH) Transport Layer Protocol.

46 IETF RFC 8040 RESTCONF Protocol.

47 O-RAN Open Fronthaul M-Plane Specification v06.00.

9

Network Deployment

Mika Aalto, Akash Dutta, Kenneth Y. Ho, Raija Lilius and Esa Metsälä

9.1 NSA and SA Deployments

9.1.1 Shared Transport

As Long-Term Evolution (LTE) is today deployed as classical base stations, classical 5G base stations typically go to the same sites with LTE and share the backhaul with it, and possibly also earlier technologies (3G and 2G). A shared site leads normally to a shared transport link too.

Cloud deployment with 5G relies on radio cloud data centres and those are expected to reside mostly on upper tiers of the network, and on larger sites – although compact far-edge sites can be pushed to access. Both classical and cloud base stations would typically connect to a common transport network, with data centre sites having networking equipment (switches and edge routers). So, since with data centre sites there are already networking services available for attaching other base stations or devices from that site to the same network, the question of shared transport is more on the classical, distributed sites and the last-mile access to those sites.

With LTE, access links were initially built with 1G interfaces. Later, 10G interfaces were also used, although capacity requirements are closer to 1G than 10G. For 2G/3G base stations, capacity requirements are even less. The 5G backhaul link rate of 10G not only suffices for the FR1 deployment of 5G, but also has room to cover other technologies in many cases. This means that common sites need a capacity upgrade to 10 Gbps, unless that is already supplied. 5G FR2 deployment with small cells adds traffic density. If those traffic flows are connected to the core network via a macro site, additional capacity may be required.

Many transport topics relate to implementation. Different radio technologies may use physically different base stations, or can also be supported on a single base station. Flows from different sources are combined by a cell-site switch or router. Switching and routing services for chaining may also be supported by base station integrated functionality. The main alternatives for combining radio technologies like LTE and 5G in the access backhaul are shown in Figure 9.1. Port rates could be higher, depending on traffic demands.

In the alternatives shown in Figure 9.1, flows from all base stations on the site are aggregated and forwarded onto the common uplink – it is just the implementation that differs. A

5G Backhaul and Fronthaul, First Edition. Edited by Esa Markus Metsälä and Juha T. T. Salmelin.
© 2023 John Wiley & Sons Ltd. Published 2023 by John Wiley & Sons Ltd.

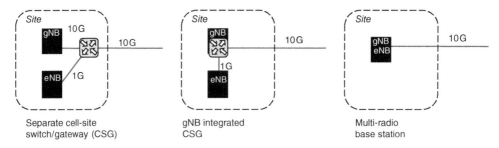

Figure 9.1 Shared backhaul.

separate Cell Site Gateway (CSG) may be used, or the same function can be integrated into the gNB. A third common alternative is the multi-radio base station, which hosts not only 5G, but also LTE and possibly 3G/2G radio functions. Flows are separated by different IP endpoints or by further information like source/destination ports, protocol, and so on. IPsec protects traffic flows in the network domain.

Use of integrated functions or multi-radio base stations minimizes the required space and cost of the site. A separate CSG, on the other hand, can provide more ports for additional (possibly non-mobile) traffic sources, further networking services like routing and IP VPNs, and also independence from the base station operation, which may help in maintaining service via another radio technology if one technology experiences an outage. This requires avoiding other common components for the site, like power supply. External CSG also provides a clear demarcation point between radio and transport operations, noting however that the IP endpoints in the base station need to be managed either by the transport or radio team.

In Figure 9.1 the access link from the cell site is assumed to be fibre, with the optical transmitter integrated into the base station or the CSG. The physical access link often requires additional devices, like microwave radio indoor/outdoor units for the wireless access link, but possibly also active or passive optical devices, and in the end the solution at the site may consist of multiple networking and transport devices. If there are external traffic sources other than mobile traffic, or other networking services are needed, then a CSG-based site solution is more likely to be required.

Connectivity to peer elements from the shared access link is arranged in many ways, depending on the network. Figure 3.6 in Chapter 3 illustrated an example where the CSG consolidates all traffic flows from the multi-radio site, then utilizes a point-to-point optical system (DWDM) to attach to an IP/MPLS network. Networking technologies discussed in Chapter 8 are examples of tools that can be used to create the required connectivity services.

The case of sharing the access link with multiple radio technologies is discussed above. Because of the cost of the access link, many other cases exist for sharing transport:

- Sharing the access link with another operator, possibly together with RAN sharing (Multi-Operator RAN – MORAN, Multi-Operator Core Network – MOCN). Different operating models are used – one of the operators responsible for operating and providing the service, a joint organization set up for the purpose, or a third party.
- Cell sites and towers are, in many networks, owned by a telecommunications infrastructure real-estate company, which leases the sites back to communication service

providers. An access link or some part of the infrastructure (like optical cabling) can be shared.

- Sharing the link with other traffic sources, an example could be the smart cities use case discussed in Chapter 5. Here, the access links carrying mobile network traffic would also carry video monitoring, for example. The network could be operated by a third-party neutral host.
- Sharing the backhaul link with fronthaul. With disaggregated 5G architecture, many 5G O-RU-based cell sites may coincide with LTE sites and sharing the access link is often the most cost-efficient approach if both traffic flows can co-exist technically without compromising the required QoS (Quality of Service) for any of the flows. Existing macro sites can also serve as hub points collecting traffic from a number of O-RUs, so where O-RU sites cannot use an existing site, some part of the access transport is still shared. Also, other (Time-Sensitive Networking) TSN-type services could utilize the same network when that provides low enough latency.

In higher network tiers where, for example, IP VPNs are supported on MPLS or other network technology, using a common network is obvious as a separate VPN service can be created for each customer and/or each use case.

9.1.2 NSA 3x Mode

5G is built with close co-existence to previous network generation (LTE) in mind, and it is not only transport that can be shared. The LTE core network is often used as the first phase of 5G introduction, with User Equipment (UE) in S1 mode and 5G–LTE dual connectivity (both 5G and LTE radio interfaces are in use).

Dual connectivity in general builds on concepts of MCG (Master Cell Group) and SCG (Secondary Cell Group) bearers, and master and secondary nodes (MN, SN). With 5G–LTE dual connectivity mode 3x, eNB is the MN and gNB is the SN. Information on 3GPP 5G dual connectivity can be found in Ref. [1]. In the user plane with MCG bearer, there is only an LTE air interface leg. With SCG split bearer, there can be both LTE and 5G air interface legs.

In the user plane, gNB has an interface to both the eNB and the LTE core network, SGW. For SCG split bearer, user data flows occur through both of these interfaces – as guided by scheduling decisions of the gNB. A direct interface from the gNB to the SGW in the LTE core enables high 5G data rates as data flow bypasses the LTE eNB, which would otherwise often be a bottleneck, due to higher capacities of 5G that do not need to go through the eNB, as shown in Figure 9.2.

The MCG bearer shown on the left-hand side utilizes only the LTE interface and the S1-U interface of the eNB is used. The mode 3x SCG split bearer utilizes both 5G and LTE air interfaces, with scheduling decided by the gNB. The split point is the (Packet Data Convergence Protocol) PDCP layer of the gNB. For the split bearer, the S1-U interface of the gNB is used.

For the SCG bearer, at one extreme all traffic may go via 5G air interface and nothing via LTE if 5G has bandwidth but LTE is congested or not available. At the other extreme, all traffic goes from the split point via X2-U to the eNB and to the LTE interface instead of 5G, which is a typical scenario when UE moves out of 5G coverage. The LTE air interface via the X2 interface is then the one carrying all traffic.

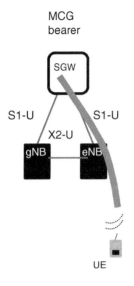

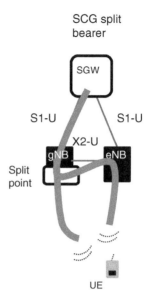

Figure 9.2 Use of 5G/LTE split bearer.

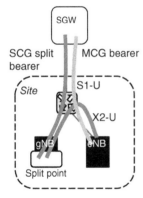

Figure 9.3 3x mode with CSG.

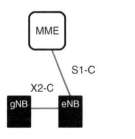

Figure 9.4 3x control plane architecture.

An implementation with an external site gateway/switch is shown in Figure 9.3.

An integrated switch/CSG can be used as well. If the 5G gNB is deployed on the existing LTE site (shared site), then the X2 interface between gNB and eNB is a site internal cable. If LTE is on a different site, X2 connectivity between the two sites is required, however the site cannot be far away since the air interfaces have to cover the same area for the UE to be able to use both.

For transport dimensioning, X2-U needs to have capacity for the maximum that the LTE air interface can carry, and S1-U of the gNB the maximum that the 5G air interface and LTE air interface together can support for the split bearer. X2-U has high traffic load, especially in cases where the SCG split bearer is established but 5G air interface conditions are poor. Radio conditions may change rapidly, so X2-U has a changing traffic pattern. At one instant the link may be lightly loaded but at the next it may be in full use, which makes the traffic flow bursty.

The split point is the gNB PDCP layer, which buffers packets if the 5G air interface is not able to deliver those to the UE and instead tries the LTE leg. The flow control procedure (in Ref. [2]) over X2-U helps the scheduler decide on which air interface to use.

For the control plane, between gNB and eNB, X2 includes a control plane interface (X2-C). There is no direct control plane interface from gNB to the Mobile Management Entity (MME) and instead the gNB relies on the LTE eNB for its control plane, as shown in Figure 9.4. LTE eNB interfaces the MME in the LTE core.

For the SCG split bearer, the control plane anchor point is the en-gNB (LTE base station). The LTE X2 interface also connects neighbour eNBs for inter-eNB X2 handover. With intersite mobility, the new (target) eNB may then again establish an SCG split bearer and the new site again supports both LTE and 5G air interface legs. For disaggregated 5G RAN architecture, the Central Unit (CU) hosts the interfaces to the eNB, the X2-U and X2-C, and also interfaces the LTE core network (SGW with S1-U). The Distributed Unit (DU) does not interface the eNB directly. For NSA mode, variations exist in 3GPP with mode 3 having the eNB responsible for the split function and mode 3a where the split function is in the LTE core.

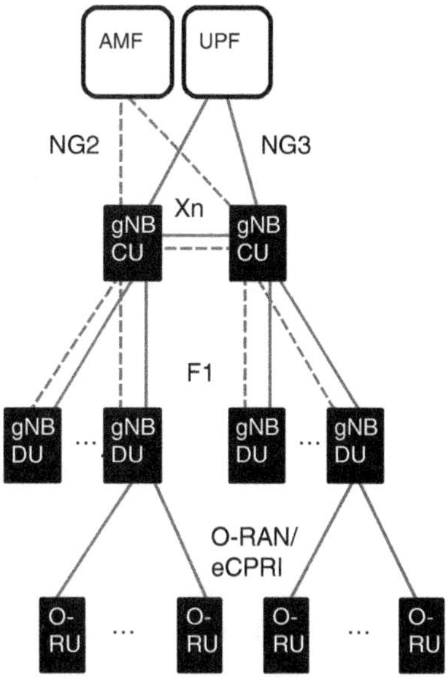

Figure 9.5 SA mode (disaggregated architecture).

9.1.3 SA Mode

In SA mode, the gNB uses a 5G core network with full 5G experience and services offered by the 5G system. NSA mode can be viewed as a transitionary step towards full 5G. SA mode applies to both classical BTS and disaggregated 5G architecture. Figure 9.5 presents the disaggregated variant, with dashed lines representing control plane and solid lines user plane interfaces.

As with NSA mode, wherever LTE or other sites exist, those would be shared for 5G and the transport link is also shared. If LTE and 5G basebands are located on the same sites, it is the F1 interface that is shared with the LTE S1 interface in the transport links. O-RUs may also be shared with the LTE baseband (eNBs), in which case 5G fronthaul (eCPRI/ORAN) is shared with LTE backhaul (S1). If LTE uses remote Remote Radio Heads (RRHs), then CPRI links are shared with eCPRI links. Dual connectivity is possible also in NR–NR mode, utilizing the Xn interface between the gNBs.

9.2 Cloud RAN Deployments

In Chapter 8 we looked at the cloud infrastructure and virtualization technologies and noted how this becomes applicable to 5G network function deployment, especially for RAN functions. Here, we see in more detail how 5G – and more specifically 5G RAN – is being deployed in the cloud.

9.2.1 Motivation for Cloud RAN

Telco operators are seriously considering and slowly shifting to the cloud RAN architecture. While operators have their own use cases depending on geographical and demographic variations in each market, some of the common drivers to consider cloud RAN are [3]:

- Business agility – enables new services and creates business opportunities through shorter software and innovation cycles, including co-hosting multiple RATs (Radio Access Technologies) efficiently.
- High pooling and scalability – allow resources to be pooled and scaled up/down efficiently based on demand.
- Total cost of ownership – this is with a long-term view on maintaining proprietary hardware (HW) inventory versus commercial off-the-shelf HW, optimized resource utilization and power usage (via high pooling, scaling and other means), OPEX savings with highly automated software delivery and integration (including the possibility to host RAN on webscalers like AWS, Azure, Google cloud and others).

Cloud RAN, while promising gains, poses some challenges in the implementation. The ecosystem is still evolving in terms of RAN-centric processors or HW accelerators, open application programming interfaces for using and managing the accelerators, fronthaul network readiness for C-RAN (centralized cloud RAN), and so on. Overall, interest is there in general, but viability in the near term varies from country to country, operator to operator, based on business model and ecosystem readiness.

Nevertheless, cloud-based deployment of 5G is a cornerstone in NGMN (Next Generation Mobile Networks alliance) 5G vision, and there is growing readiness in the cloud ecosystem to host 5G RAN functions in addition to the more naturally cloud-aligned core network functions. A coupled topic, especially from an operator viewpoint, is Open RAN (O-RAN). The whole idea is to virtualize all network functions (VRAN) and make them interoperate seamlessly, regardless of the vendor of each individual network function (CU, DU, RU). Additionally, the openness from O-RAN also pushes to make the virtualized network functions work seamlessly with the cloud platform and with any HW accelerators (L1 acceleration is the primary target here).

Figure 9.6 gives a high-level view of legacy RAN versus a virtualized RAN with a centralized vDU (virtualized DU). Note that the level of centralization of the vDU depends on the factors mentioned earlier, like the readiness and reach of the fronthaul network. Pooling and scalability are further elaborated in later sections.

Traditionally, the network functions have been deployed on vendor-specific proprietary hardware with a specific proprietary management interface. Specifically, in RAN, the hardware design is traditionally tied to Radio Access Technology (RAT) and existing equipment cannot easily host a new RAT network function, or can do so with limited capacity and performance, eventually needing an upgrade in HW equipment. With cloud deployment, the goal is to make most RAN functions agnostic to the hardware, making it easier and cost efficient to evolve to new RAT evolution features or even new RAT. This detachment of radio functions and features from the hardware is the key to this new world of software-defined RAN. Not only is this cost effective and an enabler for business agility, it also opens new service and revenue models by leveraging another key cornerstone of 5G systems – network slicing.

The deployment of core network functions naturally aligns with the readymade off-the-shelf IT cloud infrastructure, because most core network (and northward) functions are mainly data driven and based on non-time-critical transactions – not very different from typical web applications. Therefore, naturally the first 5G network function deployments on the cloud were these core network functions. However, RAN functions mainly work with air interface protocols that are sensitive to the timing of the signals exchanged

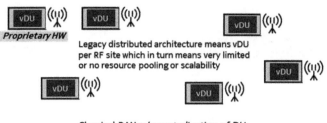

Classical RAN w/o centralization of DU

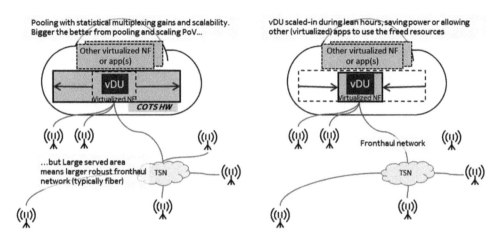

Centralized Cloud RAN with 'big'
centralized vDU

Figure 9.6 Legacy RAN vs centralized cloud RAN.

with the mobile equipment, and therefore most RAN protocol layers impose soft real-time
or hard real-time requirements on the underlying platform for those network functions.

As we can see from Figure 9.7, as we go from the top down the protocol stack of each
network element (or network function), the timing sensitivity increases and the throughput
on the interface also increases. Therefore, the most easily cloudified 5G RAN function is
CU-C/CU-U, because interfaces of CU are not time critical. CU-C/CU-U are deployed as
independent VNFs, on the same or different data centres, interworking via the 3GPP standardized E1 interface. There can be multiple CU-Us served by a single CU-C, and there can
be multiple (not many in practice) DUs served by a CU-C + CU-U combination. CU-C
anchors the call/mobility control (and the RRC protocol) of the UEs and terminates the
control plane link to the core network (NG-C) and other interworking RAN elements
(eNBs, gNBs, DUs), while CU-U handles the data path and terminates the user plane links
to the core network (NG-U) and other RAN elements (eNBs, gNBs, DUs). The DU element
does most of the air interface lower-layer processing and has soft and hard real-time protocol functions. There is active and aggressive work ongoing to virtualize the DU as well,
although it has some near-hard real-time functions. Virtualization of the DU needs some
HW acceleration support for some Layer1 functions to make it more commercially viable.

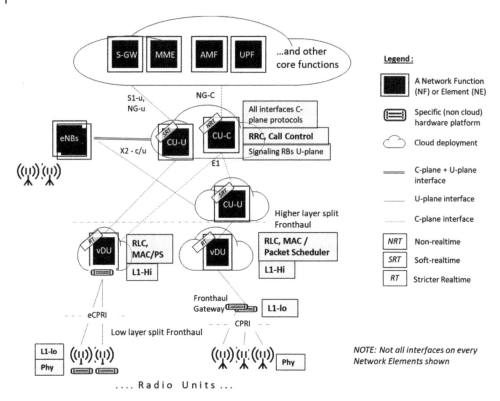

Figure 9.7 Key protocol layers and workloads in RAN network functions.

We shall see more on the DU virtualization in later sections. By separating the control plane and user plane of CU and splitting the protocol layer processing between CU and DU, 3GPP enables flexible deployment of CU-C, CU-U and DUs geographically, based on the Telco operator's business case, coverage and service plans. The CU-DU split is a bit like the UMTS RCN-NodeB split, but the key improvement in 5G fronthaul higher layer split (HLS, the CU–DU interface, aka F1 interface) is that it is very delay tolerant because of a clean protocol cut at PDCP-RLC layers. This gives flexibility and convenience to transport planning – enabling very flexible placement of CU-C, CU-Us and DUs of a gNB.

9.2.2 Pooling and Scalability in CU

In a cloud RAN deployment scenario, the CU-C VNF is typically deployed in the central office location for a geographical area like a region or big city and its suburbs. This enables large pooling gains on resource utilization for the C-plane functions and it is also easy to enable transport for low-throughput medium-latency F1-C and E1 interfaces from the CU-C location to the location of CU-Us or DUs (order of tens of ms acceptable delay). For example, a region with population of about 1 million in about 100 km^2 (urban area) can have hundreds of DUs with up to 16 000 cells in total and not every DU or cell is equally loaded at any point in time. Office areas peak during office hours, while residential areas and shopping areas pick up in the evenings, meaning that the load offered from cell sites (via the respective serving DUs) varies over time based on geographical area. So, it makes sense to under-dimension the CU-C to take advantage of this statistical gain when a large pool of users is served by a single CU-C. A consideration of 50–70% statistical gain is common if the pool size is large, which in turn delivers savings in hardware infrastructure, energy and transport costs. Where cloud deployment brings benefit in such cases is in the scalability and elasticity of the network functions, like CU-C. At peak load hours the CU-C can scale up to provide more capacity, while at off-peak night hours it can scale down (and potentially some other backend application or different time zone application can scale up to use the freed cloud capacity if needed, or energy savings can be applied on the under-utilized resources).

A CU-U can similarly be deployed as a VNF in the central office location and provide a large pool to make use of pooling gains and elasticity of the CU-U VNF. This works well with regular internal or eMBB services, where the latency requirements (on F1-U and X2/Xn-U interfaces) are not too strict (order of tens of ms latency). Other options to deploy CU-Us are at aggregated edge sites (like BBU hub or hotel) or even at radio sites, depending on the business case. Deployment farther away from the central office means smaller CU-U VNF deployments on mini clouds or micro clouds (on the so-called far-edge data centres), giving a smaller pooling gain but a better latency for enabling, for example, some low-latency services (by also co-deploying the UPF function and optionally some MEC functions with the CU-U, e.g. for URLLC services).

Pooling in the data plane is more important as it takes more resources in the data centre as well as in the transport network. The implementation of pooling and scalability in the CU is relatively simple, as the processing and transport usage is only data bearer (of the mobile user/s) centric. In Figure 9.8, the simplest design is that users/data bearers load is shared across the N active servers and the number of such active servers can be scaled in and out based on the traffic load. The scaling of external endpoints (e.g. F1-U endpoint) is also natural, as the data plane endpoints are negotiated over the control plane procedure during the call/leg setup between peers. The only challenge is in handling IPsec tunnels while dynamically scaling the data plane resources. 3GPP improvements have now covered these original drawbacks by allowing peers to also exchange IPsec endpoint address during the call setup procedure. In the absence of this, one may need to have out-of-band operability-driven pre-configuration in peers to allow dynamic scaling with IPsec, or have a separate Security GW outside the CU.

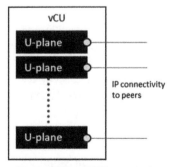
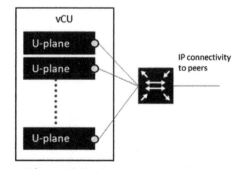

Direct visibility of IP/IPsec endpoint to peers: scaling directly visible to peers

IP/IPsec endpoint abstracts the internal U-plane processing to peers: scaling not visible to peers

Figure 9.8 Capacity scaling with or without external endpoint scaling.

9.2.3 High Availability in CU

This section talks about high availability and service resilience from the CU standpoint, though the principles are applicable to any application, including inside the vDU, at least in part.

The CU typically being a centralized entity and serving a large geographical area and many users, one of the key service level agreements is service availability. Availability is the ability of the system to provide its service without failing over a designated period. The typical availability target for a Telco system is 99.999% (aka five nines i.e. 5-9s), meaning it should not be unavailable for more than about 5 minutes over an entire year (or about 26 seconds in a month). This is a challenging reliability target, especially in the cloud, because the cloud infra has a typical availability of 99,99%. If a server fails, someone detects the failure – and perhaps tries to restart – before finally going for a swap of the server. All of this can take several tens of minutes to several hours, creating a pro rata downtime of capacity long enough to bring down the overall availability. This is where the importance of application service resilience and redundancy models comes into play. Figure 9.9 illustrates failure detection.

Service resilience generally means that a service, after being restarted/recovered from any kind of failure, should be able to continue from where it failed, without causing other services to stop or malfunction. Service resilience in the cloud is usually implemented via some level/variant of the stateless software design paradigm. Service resilience is not a new concept, but with cloud deployment it can be done more easily due to the variety of ready-made enablers out there in the ecosystem.

Another aspect of high availability is the redundancy model (see Figure 9.10). 2N, N + M and N+ are common redundancy models in general, though a truly cloud-native design has stateless N+ software components (as microservices). N+ model means that there are N service instances which share the load, distributed via a traffic dispatcher or load balanced. there may be some load headroom in each of these N instances so that if any of them fails, its load can be re-distributed to remaining instances. Of course the faulty instance is expected to come back after appropriate recovery procedure. It is a matter of effort versus benefits in taking design decisions about the redundancy model for each application service or SW component. For example, some components can be 2N redundant (hot or warm or cold standby) to favour fast and predictable service recovery, while others can be N+ load sharing (and scalable). Kubernetes by default only recognizes the N+ model, so some special application-level logic is usually needed to emulate 2N redundancy in case the CNF is deployed on a Kubernetes cluster.

Service resilience and an appropriate redundancy model mitigate the low infrastructure-level availability and enable

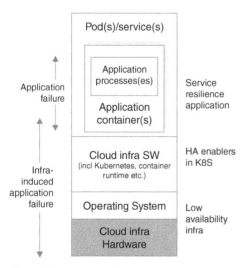

Figure 9.9 Failure directions and enablers in container based cloud deployment.

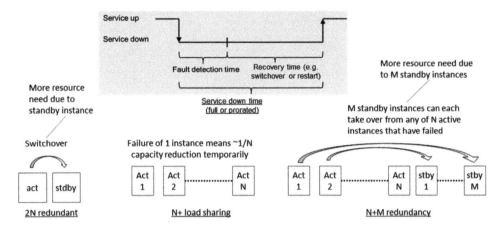

Figure 9.10 Redundancy models and downtime.

application to still meet the target of 99,999% or even 99,9999% availability. One more aspect to note is that the availability itself still does not implicitly clarify whether the existing calls are maintained or not. Service resilience in its basic form can be just to handle service or infra failures by not impacting external interfaces (F1, X2, etc.) – not letting peers see a permanent or semi-permanent failure of the network element that has experienced some internal service or infra failure, existing calls being released but new calls (including re-attempts of those failed calls) immediately being serviced. The ultimate level of service resilience is that existing calls just experience a glitch, but most of them can continue without requiring a re-attempt.

The third aspect of overall high availability is network redundancy. A typical data centre connectivity blueprint has redundant switches and sometimes redundant NICs on the servers as well (refer to Chapter 8). CU external interfaces are non-real-time IP interfaces, which makes the link redundancy design much simpler compared to the vDU. Link redundancy protects against port or NIC or switch failures, apart from cable cuts in the path to the peer network elements. A bonding interface with Link Aggregation Group (LAG) is a common solution for active–active redundancy, but other solutions for active–standby or active–active redundancy can also be employed – especially with SRIOV-based interfaces. In the case of active–active link redundancy with LAG, sometimes an overbooking strategy is applied to optimize the HW solution. For example, if the maximum bandwidth (BW) needed for a server is 40Gbit/s, then link aggregation can be created over two 25GE interfaces giving a total ~50Gbit/s pipe when both links are working (subject to enough entropy in the connections so that both links get utilized fairly with the hash function in use for LAG). If any of the paired links fail, then the total available pipe BW drops to 25Gbit/s, resulting in a KPI and capacity drop until the failed link is repaired. Irrespective of the CU application-level resilience needs, the cloud infrastructure itself relies on link redundancy within the data centre to have reliable control and monitoring data exchange among the nodes in the cloud cluster.

System high availability must also consider site disaster and therefore geo-redundancy is important. Geo-redundancy means a standby system available at another geographical location to cover for catastrophes at the primary geo-location of the CU (see Figure 9.11).

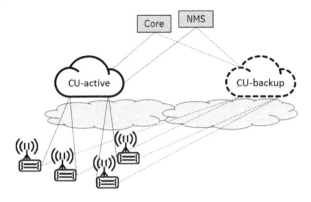

Figure 9.11 Geo-redundant CU.

The solution is conceptually about the whole data centre-level redundancy, which means all NF deployed on the failed data centre will be backed by NF at a remote data centre. The latest configuration – and potentially even the call states – can be transferred from active to standby NF across the data centres. The core network and NMS/EMS see the CU 'move' from one place to another (the gNB ID and cell IDs, etc. are all retained).

From the transport perspective this means that the back-up data centre will have enough backhaul capacity to cater for the CU popping up in case of primary failure. It also means that all DUs, eNBs (in the NSA case) and neighbour gNBs served by the CU should have the necessary connectivity and link capacity towards both data centre locations. The same applies if CU-Us are distributed across the geography, then each CU-U should be able to work with the back-up CU-C in case of geo-switchover.

A more advanced version of geo-redundancy is N:M redundancy, meaning there are M back-up CUs for N different active CUs. The distribution of back-up CUs can be flexible across the available data centres. This saves resources by avoiding a 1:1 back-up solution for each active CU. One way of preserving resource in the back-up CU is to have it stay at the lowest configuration and size while it is acting as back-up, and scale it up after it becomes active to meet the traffic load.

9.2.4 Evolving to Real-Time Cloud – vDU

Just like the CU-CU deployment options, the DU or vDU is also deployable in multiple ways at various levels of aggregation and resulting pooling gains. A vDU deployed at the cell site would usually be of small enough (as small as a single server) capacity to serve the cell site (a few FDD cells and/or a few TDD cells). Such a vDU deployment is also called cloud D-RAN deployment (distributed RAN). Such a small vDU can be viewed more like an appliance rather than really a cloud network function – it would usually be neither scalable nor highly available. A more cloud-like deployment of the vDU would be C-RAN (centralized cloud RAN), where the vDU is much bigger and serves multiple cell sites. The only challenge in C-RAN deployment is that the connectivity to the remote cell sites must be good enough to give the required latency for the fronthaul link to the radio units on those cell sites (recall the beginning of this chapter). While the F1 link can tolerate multiple tens of milliseconds latency, the fronthaul link (DU ↔ RU) latency tolerance is of order 100 μs. This impacts how many cell sites (in a geographical radius) can be connected to a C-RAN hub site having the vDU. Additionally, such a 'C-RAN' vDU must adjust its fronthaul delay handling on a per radio unit basis, as it has radio units connected over a variety of distances. A large vDU deployed across multiple servers (e.g. tens of servers) creates a large pool of processing resources for the traffic offered by the hundreds of cells served by such a C-RAN vDU. The advantages are pooling, scalability (and resulting energy saving), service resilience and overall better Total Cost of Ownership (TCO) than a D-RAN installation. A note about scalability is that there is a similar choice of design as vCU for managing the F1-U endpoints (IP addresses) – directly visible to peers or via a load-balancer or front end.

Distributed and centralized vDUs are illustrated in Figure 9.12.

9.2.4.1 Pooling and Scalability in the vDU

First, we look at general pooling and scaling concepts in the context of C-RAN vDU. Pooling mechanisms aim to extract statistical multiplexing gains from uneven distribution of load across the service area (collection of cells) of a vDU. Pooling solutions can be coarsely categorized as intra-server and inter-server, within the vDU instance. The below explanations are mainly w.r.t. user plane processing, but the same concepts work for any other kind of application in general.

As noted earlier, one key advantage of C-RAN vDU is pooling and scalability, which directly impacts the TCO by maximizing average cell density (cells per server), data centre energy consumption and fronthaul bandwidth requirements. On a certain server, there can be big differences in long-term TCO if there are, say, three cells with full independent capacities or six cells with a pooled capacity. As was noted earlier, not all cells peak in traffic at the same time. Pooling of capacity (number of connected users, number of UE/TTI, MIMO layers, number of scheduled PRBs, etc.) across a set of cells within a server is a very effective way of TCO optimization. It not only reduces the need for HW resources, but also reduces the BW on fronthaul. This in turn means less switching capacity on the fronthaul network, less equipment overall and less power consumption overall. Such intra-server pooling (illustrated in Figure 9.13) is applicable to both D-RAN and C-RAN vDU deployment cases.

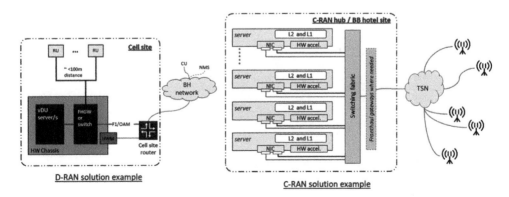

Figure 9.12 Distributed and centralized vDU deployment.

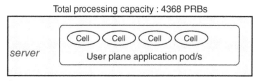

No pooling – resources reserved for 4 cells, all peaking together
(not likely in practice)

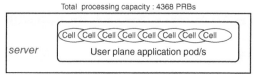

2:1 pooling – resources reserved for 8 cells, max 4 peaking together
(practical assumption on real life traffic pattern)

Figure 9.13 Basic server-level pooling.

Typically, such server-level pooling is configurable to suit the needs of particular use cases.

On top of such intra-server pooling, there can be inter-server pooling (as in Figure 9.14), which works across the servers at the far-edge data centre with one or several vDUs deployed there. The most rudimentary pooling solution at a C-RAN hub site has multiple small vDUs (say one vDU per server) and RUs are rehomed administratively across those vDUs when needed (optimization or failure recovery cases), see Figure 9.15.

However, since this involves rehoming of RUs across vDUs and works with the rules of 3GPP-based procedures, it is slow and has inherent limitations (e.g. the whole RU must talk to a single vDU, if the RU has multiple cells then all cells must go with the RU).

Such pooling via RU rehoming can be done much better if the C-RAN site has a bigger and single vDU (short name C-RAN vDU). Here the load redistribution can be done even at cell level, because all RUs are handled by the same vDU (as in Figure 9.15).

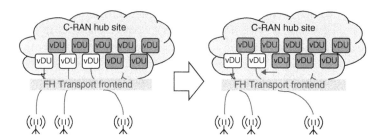

Example: during lean hour, a selected RU re-homed and
one more vDU is turned off

Figure 9.14 Very basic inter-vDU pooling at C-RAN site.

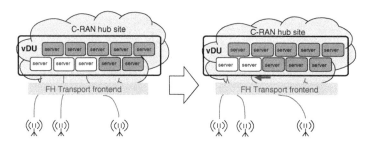

Example: during lean hour, a selected RU/s or cells re-homed
across servers in the vDU, servers can be turned off then

Figure 9.15 Inter-server pooling (within a vDU) at C-RAN site.

O-RAN has also defined the above forms of baseband pooling; extreme pooling is where traffic across cells should be balanced and remapped fluidly across servers within the O-DU or set of O-DUs at that far-edge site. This requires either a common front end for fronthaul transport with the cell and user functions of L2 and L1 distributed across servers of the C-RAN vDU, or eCPRI streams for the same RU terminating on multiple vDU servers with users distributed across those streams.

Scalability is closely related to pooling. If the above scenarios are executed in runtime, then there are opportunities for scaling the resources as well. For example, when all cells or RUs are moved away from the BB processing pods, then those pods can be scaled in (removed), thereby relinquishing virtualized HW resources from the server where the pods were hosted. On other hand, when there is a need for more processing, new BB processing pods can be scaled out (added) and then some cells/RUs can be rehomed to those new pods to redistribute the load within the vDU. This is the general scaling concept and not limited to BB processing of cells alone, although it is used as an example here. If the extreme pooling mentioned in the previous paragraph is implemented, it gives the benefit of flexible and fine-grained scaling to match the load in terms of number of cells, number of users and throughput.

9.2.4.2 Fronthaul Dimensioning Considerations

Fronthaul is the link between DU (or vDU) and the RU (Radio Unit) network elements; focusing on the eCPRI interface (LLS 7.2), it carries the frequency domain IQ data between those network elements. The amount of traffic on the fronthaul interface is driven by the cell fronthaul split type (LLS option 7.2 or 7.3), cell bandwidth, cell type and numerology (TDD/FDD, SCS), number of special streams, compression enablement and the RRM and pooling features active on the DU and RU. Apart from the IQ data, fronthaul also carries the control plane (or Fast Control Plane, FCP) messages.

O-RAN standardizes all control, user, synchronization and management planes (C/U/S/M-planes) of the fronthaul interface between the O-DU and O-RU (O-RAN terms for DU and RU). The FCP does not take much bandwidth, but it is important to help regulate the U-plane BW consumed on the fronthaul interface. FCP basically tells the O-RU what is coming on the U-plane and can, for example, help control fronthaul BW usage by sending only the scheduled PRBs (Physical Resource Blocks) on the DL (and similarly on the UL). IQ data compression is another important factor, as listed earlier, since it can do an envelope compression of 16-bit words to 9- or 7-bit words, thus reducing the required bandwidth by a factor of over 50%. Another important factor in O-RAN fronthaul bandwidth requirement is whether the O-RU is category A or category B (ref O-RAN.WG4. CUS.0-v10.00). Other evolving alternatives are LLS 7-3, and 3GPP LLS option 6.

Server-level pooling algorithms also help optimize fronthaul data rates. If a server hosts, say, 12 cells, the fronthaul link from that server can be under-dimensioned to carry peak fronthaul traffic for three of them at full load and the other nine cells at 50% load. This means at any time the cumulative fronthaul data rate is equivalent to 7.5 peak cells. This is enabled by pooling features like layers pooling or PRB pooling, as these pooling features work essentially in the packet schedulers by limiting the cumulative fronthaul data rate of the server.

9.2.4.3 Server and NIC Dimensioning

One key TCO parameter for the vDU is cell density, or number of cells per server. This can be calculated with different perspectives and depends on the software design, deployment use case of the vDU and HW platform, but a rough measure comes from the number of cells that can be processed on a worker node in the Kubernetes cluster hosting a vDU alone. The processing needs are mainly for the user plane functions – L2hi, packet scheduler, MAC layer, L1hi and eCPRI termination function. Today, the predominant design requires all user plane functions – except parts of L1 and optionally some eCPRI functions (that are offloaded to the HW accelerator) – to be running on the main CPU cores. Depending on the HW accelerator, the offloaded parts of L1 can be at least the FEC/LDPC codec and can go all the way up to complete L1 function running on the HW accelerator. Some smart NICs (or NICs integrated into the HW accelerator card) can also run eCPRI functions like HW-based scheduling of uplink eCPRI data to the target cores, or provide complete eCPRI termination including IQ data compression. This topic was explained a bit more in Chapter 8. While today the HW accelerators are mainly PCI-based add-on cards, next-generation processors are coming with more and more HW acceleration built in, reducing the need to have extra PCI add-on accelerator cards, at least for lookaside acceleration cases.

It is important that the server has a more or less balanced capacity in terms of HW accelerator capacity, CPU core capacity, NIC capacity and memory. All these must be carefully dimensioned to reach a cost-efficient and power-optimized server profile. Many of today's NICs already have 1× or 2× QSFP, giving 2×100G or 2× 4×25G lanes. In a cloud RAN solution, the preferred fronthaul connectivity is switched fronthaul, meaning that the fronthaul eCPRI ports on the vDU servers connect to a switch and the RUs are connected (directly or via TSN) on the other side of the switch. This gives a full mesh connectivity of any RU to any server. The QSFP ports on the server NIC can be used as 4×25G or 100G aggregated link. Obviously, having N×25G individual lanes creates the BW fragmentation problem, while having a single 100G fronthaul gives better efficiency of packing more antenna carriers into that same pipe. This may need multiplexing of infrastructure traffic on the same aggregate 100G pipe that is being used for fronthaul eCPRI and PTP traffic, which requires careful consideration of QoS maintenance for fronthaul and sync traffic. Note that the NIC ports in the context of fronthaul interface are part of the inline L1 accelerator, and potentially in separate NICs if a lookaside L1 accelerator is used in the vDU.

9.2.4.4 High Availability and Service Resilience

For C-RAN vDU, service resilience is important as it will service multiple tens or potentially hundreds of cells. Service resilience for C-plane, F1-C, operability functions is like that of vCU – as those functions are not handling the fronthaul interface or data plane. For such non-real-time state functions, 2N redundancy can be deployed or an N+ load-sharing redundancy scheme can be used for the stateless services. For the L2 packet scheduler or L1, is it trickier to have complete service resilience as there is real-time context data to handle and replication of such data is difficult. An easier way is to have smaller N+ redundancy so that failure of an instance does not impact the whole vDU service but only a subset of cells, for a short period before recovery.

9.2.4.5 Link Redundancy Considerations

An additional factor for vDU fronthaul is redundancy considerations. While link redundancy is important for the overall availability of the network functions, the cost of E2E link redundancy can be high and that can indirectly be a driver for sizing the vDU so that it is not too big and its link failure (through a port, NIC, switch failure or cable cut, etc.) does not cause substantial network outage. In Figure 9.16, for a C-RAN kind of deployment with a vDU hosted across multiple servers, the most important redundancy is that for the cloud infrastructure – because failure of infra-level connectivity to a server (the controller or master nodes should always be able to talk to the compute or worker nodes) immediately takes that server out of the cloud (e.g. the Kubernetes cluster).

The next important link is for the vDU midhaul carrying F1-C traffic, because a break in that link impacts the whole vDU service. F1-u is also critical, but typically there can be multiple F1-u links emanating from different servers (for U-plane scalability), so failure of a single link has limited impact. Of course, if F1-u is aggregated to a single link then the redundancy of that single F1-u link should be enabled.

Fronthaul is almost certainly distributed across the servers, for example each server hosting a certain number of cells that emanates an individual fronthaul eCPRI or CPRI link towards the remote RUs on the cell sites. Redundancy on the fronthaul can be slightly tricky due to the low tolerance to link break (during failure detection and recovery or switchover), especially to maintain the cell service during recovery. Nevertheless, link and switch redundancy are still beneficial in large C-RAN vDU cases due to the level of impact that a switch failure in the path can cause for RAN availability. As said earlier, such path and switch redundancy are costlier as more fibre needs to be laid on the fronthaul network to create a dual path (e.g. a ring topology) to each radio site from the C-RAN vDU location.

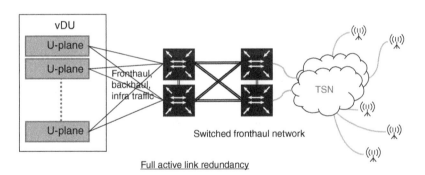

Full active link redundancy

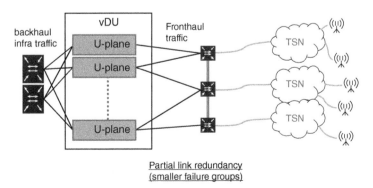

Partial link redundancy
(smaller failure groups)

Figure 9.16 Fronthaul link redundancy.

An alternative solution is to have smaller failure groups (groups of cell sites connected via a path and switch infrastructure) so that any single switch or link failure has limited impact and the rest of the cell sites are still able to provide the access service to the users, as in Figure 9.16. This can be coupled with careful radio planning, so that some radio sites can give larger macro cells overlapping with a set of smaller (throughput focused) cells so that users can find the overlapping alternate cells in case of failure of their connected cells. This can also be across RAT, for example LTE or eGPRS cells can cover for multiple 5G cells.

Today, most of the fronthaul eCPRI is L2 based (lacking IP stack), so an IP-based solution is not feasible and Ethernet link monitoring-based switchover is needed for link redundancy. The eCPRI fronthaul is evolving towards L3/IP, which can make the fronthaul link and endpoint management much simpler overall compared to the L2-based fronthaul, but the IP stack processing and BW overhead will be additional.

9.2.5 Enterprise/Private Wireless

A growing market for 5G RAN is enterprise or private wireless networks. The use cases and deployment options are described briefly in Chapter 8. The enterprise VRAN use case introduces (but is not limited to) a scenario of hosting a small (cloudified) gNB on the enterprise IT cloud, catering to a very small number of cells (e.g. one to three) as in Figure 9.17.

It could also collocate core functions to give a direct internet connectivity from the enterprise private 5G network (like a replacement of Wi-Fi). This means deployment of vCU and vDU on a very small footprint (like a single server). Such enterprise/private wireless use cases can be met easily with a downsized deployment of the cloud RAN solution along with the UPF and other core functions and potentially some MEC kind of functions, all virtualized and deployed on the same cloud infrastructure, all managed and orchestrated with a common management system. This homogenization of Telco service network functions not only brings benefits such as cheaper IT infrastructure (meaning low proprietary HW inventory) but also high automation opportunities.

In such scenarios, the HW infrastructure can be further optimized by having a HW accelerator PCIe card to offload L1 processing from the main CPU, thus giving enough room for RAN and core applications on a small footprint.

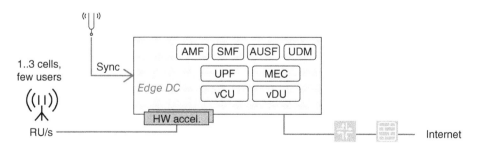

Figure 9.17 Example of enterprise RAN.

9.3 Fronthaul Deployment

9.3.1 Site Solutions and Fronthaul

LTE supports remote antenna units using CPRI framing, carrying digitized time-domain samples over direct fibre. CPRI is an industry-driven initiative [4], which standardizes the interface between the RF unit and baseband, with CPRI terminology of RE (Radio Equipment) and REC (Radio Equipment Control). The interface is not packet based but rather a Time-Division Multiplexed (TDM) bit stream with specified framing. OBSAI has defined a similar type of interface.

Before that, analogue RF modules (remote RF heads) were utilized, with RF cables connecting the baseband to the RF amplifier and antenna. Actual antennas might be at some short distance from the RF module, with antenna cable in between, with loss of power due to the cable and mast head amplifiers possibly used.

So, many kinds of site solutions have existed and continue to be used, with different types of site internal cabling between elements and modules, most of which are not considered transport or networking. The site type immediately suggests a certain configuration with practical considerations like installation type, space available, weight of equipment, height of the installation, and so on. Simple remote equipment is easy to operate and maintain, but may otherwise have limited capabilities.

With digitalization of the interface with CPRI, the cabling that was previously considered as part of the site solution started to have characteristics like digital signals, optical fibre, site-to-site connectivity, longer distances and support for chaining. A baseband hotel site connects to the remote RUs over fibre links using CPRI, with distances up to 20–40 km, typically over point-to-point fibres (although other optical technologies were also used), but no networking services like switching or routing.

With eCPRI the major step occurred with the introduction of Ethernet to the fronthaul links between RUs and baseband. Ethernet is clearly on the networking side of technologies as opposed to CPRI. With 5G a new functional split between the RF unit and the baseband had to be defined, as otherwise remote RUs would have required huge bandwidths with the split-8 option. Another significant change is that with massive (Multiple Input Multiple Output) MIMO and beamforming, the antenna can be integrated within the RU, becoming an active antenna. This means that at the site, the eCPRI interface extends to the RU with combined antenna/RU (O-RU) as in Figure 9.18.

In Figure 9.18 an example site is depicted equipped with three FR1 O-RUs. One fibre pair (Tx, Rx, unless bi-directional transmitters used) is required for each of the O-RUs, so altogether three fibre pairs for the site, with eCPRI protocol over Ethernet or IP on the fibres. The site picture is simplified, requiring also some power supply that is not shown.

The baseband (O-DU) may be located at the same site, which makes the baseband network distributed as is often done with LTE, or the baseband may be centralized to some 20 km distance. The fibre count (3× fibre as shown per site) can be reduced by optical multiplexing or a packet-level switching service. Either method adds some equipment, optical or switch, to the site.

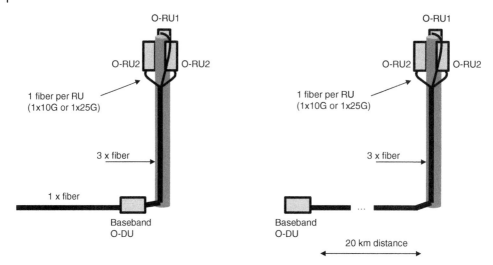

Figure 9.18 Site example.

While fronthaul is often assumed to have fibre-based access with the methods discussed in Chapter 6, likely for the majority of all sites for fronthaul connectivity, high-bandwidth wireless transport (tens of Gbps in E and D bands) is an alternative, as discussed in Chapter 7.

9.3.2 Carrying CPRI over Packet Fronthaul

Evolution in fronthaul to Ethernet and IP networking means that new deployments rely on those technlogies while the existing network is built with CPRI links. The optical layer can carry both CPRI and eCPRI by dedicating a fibre or λ for each, and in this way supports a common optical layer network providing bandwidth for both technologies.

Another alternative for sharing the fronthaul with CPRI is to map CPRI over Ethernet transparently (or with some optimizations), or convert CPRI to eCPRI. The benefit is that when CPRI is also in Ethernet frame format, this flow can be combined and switched with eCPRI Ethernet flows, using Ethernet multiplexing or switching services. The IEEE 802.1CM profile also includes definitions for the CPRI traffic flow. Latency, loss and other characteristics have to be guaranteed for both the eCPRI and CPRI flows over the packet network.

IEEE P1914.3 [5] includes two alternatives for Radio over Ethernet (RoE): structure-unaware mapping taking essentially the CPRI flow and simply applying Ethernet framing and structure-aware mapping, taking into account different CPRI flows in mapping. With RoE, existing CPRI-based RUs are merged into a new common packet network that is built for the new Ethernet-based eCPRI O-RAN RUs. RoE essentially emulates CPRI connection over an Ethernet service, as otherwise the CPRI-based RUs would have to continue to be served by dedicated CPRI links. The RoE specification has been enhanced to also cover mapping to UDP/IP.

Another approach is converting CPRI to eCPRI. The network node to carry this out is commonly called FHGW (Fronthaul Gateway). O-RAN has created a reference design specification for the hardware [6].

The use of RoE, or CPRI-eCPRI conversion, allows cases like depicted in Figure 9.19. In the upper part the RoE application and in the lower part the conversion to eCPRI is shown.

The main difference is that in the conversion to eCPRI in FHGW, radio-layer processing is also done and the split point converted, for example, from split 8 to split 7.2, so the bit rate required is reduced. The device (switch with RoE support or FHGW) is located close to the RUs (at the RU site) or to some intermediate site in the network, which then needs only Ethernet connectivity service. At the cell site, the switch or FHGW can, for example, be part of the site solution.

The network in Figure 9.19 is shown as two segments: the first segment where both eCPRI and CPRI are used, and then the segment where all traffic is Ethernet flows. In the latter segment, only Ethernet connectivity is needed, and this allows a converged network layer for existing and new RUs, with a full set of services for resilience, recovery and dual attachment.

9.3.3 Statistical Multiplexing Gain

Point-to-point fibre links all the way from the RU to the baseband are staightforward conceptually and have been a common way for realizing connectivity from RUs to centralized baseband hotel sites. With 5G this approach is possible, but requires a fibre-rich environment where each RU can have at least one dedicated fibre or λ. The drawback is obviously the huge number of fibre links. It also requires a lot of ports at the baseband end. If this can be economically arranged, it suits the real-time nature of the fronthaul traffic flows well.

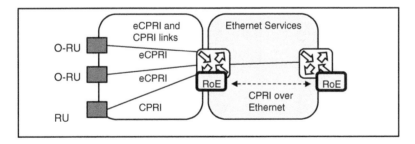

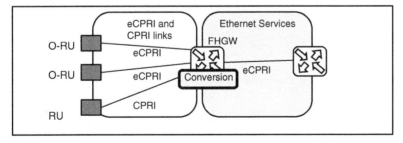

Figure 9.19 Carrying CPRI over Ethernet fronthaul.

One of the benefits with eCPRI and Ethernet and IP-based fronthaul is the possibility for statistical multiplexing. When many RU sites connect to a hub site, it is rare that all of the RUs experience peak traffic requirement at the same time. During peak traffic (busy hour), the sum of actual traffic from all RUs is very likely somewhat less than the sum of the theoretical maximum each RU can source. With eCPRI, there is additional gain in bandwidth since with eCPRI there is no need to transmit user plane data samples during an idle air interface. An example is illustrated in Figure 9.20.

In the example case of Figure 9.20 each site has three O-RUs and those are multiplexed optically to one link towards the switch. The switch connects five cell sites each with a similar configuration, so there are 15 O-RUs as traffic sources. The diagram on the left shows an example of the bandwidth needs of each of the three O-RUs at site 1 at different time instants, where the traffic amount is varying but adding the traffic sources together does not occupy a bandwidth up to three times the O-RU peak capacity, although this could also happen with some (depending on assumptions, possibly rather low) probability. If this is accepted, statistical multiplexing gain can be exploited by switching where the switch is – directly at the cell site or on a remote hub site. The statistical multiplexing gain increases when the number of independent sources increases, so the situation improves when there are at least 15 O-RUs for example, and not just three, which is still a small number. Very much depends on the traffic volume assumptions at each site during the busy hour, and to what extent the traffic sources correlate. Related pooling gains in the cloud are possible, as discussed.

Fronthaul is real-time sensitive and assumes a connectivity service of almost zero congestion and loss, which limits statistical multiplexing possibilities, so a balance needs to be set with the conflicting requirements for zero congestion/loss and excessive overdimensioning. Even when dimensioning for network-level statistical multiplexing gains, this should not compromise the air interface performance, a topic which was discussed in Chapter 4.

At the other end (baseband, vDU), fewer ports are needed when the traffic is ingress multiplexed prior to the baseband (vDU), with only a few high-capacity ports. For resilience, the attachment is duplicated.

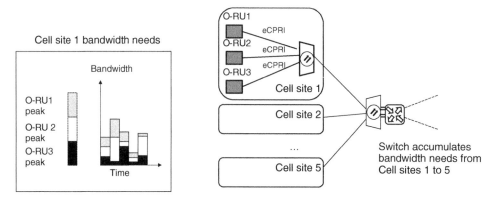

Figure 9.20 Statistical multiplexing gain.

9.3.4 Merged Backhaul and Fronthaul

For fronthaul with eCPRI, connectivity alternatives rely on any of the possible technologies in optical and networking layers. With the Ethernet protocol version of the eCPRI, an Ethernet connectivity service is needed. This can rely on direct dedicated fibres, as with CPRI. The shortcomings listed in the previous section suggest network-layer services for efficient multiplexing and related capacity and port count savings in the network, and also networking services for resiliency.

For combining traffic flows, fronthaul has clearly stricter latency and loss requirements than backhaul, and these need to be supported. With the evolution of the network, addition of RUs and small cells, network topologies other than simple point-to-point to baseband become increasingly more common. This also introduces more alternative path options which are beneficial for availability of the transport network and for reliability of the service delivery.

Fronthaul deployment assumes at least disaggregation of the gNB into O-RU and DU/CU, but possibly also further into DU and CU. Disaggregation of the RAN into remote O-RUs can be done without deploying radio clouds of any type, yet since one of the targets of remote O-RUs is to centralize other functions (baseband computing), cloud deployment for DU and CU is often assumed. With remote O-RUs, it is a benefit if the fronthaul network also allows cloud deployment, in an initial phase or later.

In Figure 9.21, different types of sites are shown with connectivity needs illustrated by dashed lines and types of logical interfaces listed on the sites. O-RUs here refer to eCPRI interfaces, while RU means LTE with CPRI. Due to cost reasons, access links are preferably shared so that all source flows of traffic from a site use the same uplink(s). Connectivity to

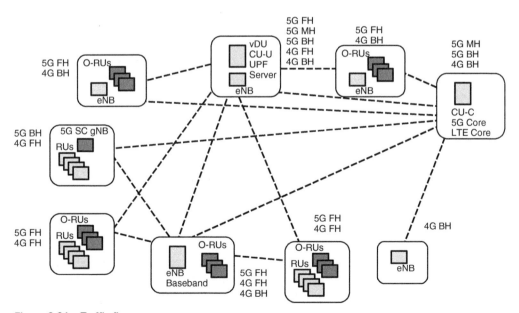

Figure 9.21 Traffic flows.

the peer element is arranged by network services in between, including dual homing and resilience.

Analysing the case, the connectivity needs of the sites are varied, with combinations of 4G and 5G FH, 4G BH and 5G FH, vice versa with 5G BH and 4G FH, and also both 4G and 5G FH with 4G FH. The vDU site needs MH (F1 interface), with 5G FH/BH and 4G BH. The diagram is simplified as, for example, the intra-RAN interfaces (Xn, X2) are not shown. Many sites depend on both fronthaul and backhaul interface connectivity.

Currently, fronthaul is often on its own network entirely if there is a dedicated fibre or λ per RU. A backhaul link, or another fronthaul link, could be added there as well by allocating another fibre so that essentially fronthaul and backhaul are on separate optical domains. Continuation of this separated approach includes adding a switching service for fronthaul bandwidth savings, while still logically managing the fronthaul separately from backhaul and midhaul.

Growth of the fronthaul network in terms of nodes and functionalities leads easily to a case where an evolution to switching in fronthaul for multiplexing gains alone is not adequate as requirements for resilience become more important and services for managing alternate paths and failure cases are needed. This can again be arranged in many ways, but in the end the fronthaul network starts to require the same services as the backhaul – or almost any larger packet network for that matter (of course with an additional consideration for low latency and low loss for the fronthaul links and nodes).

A further driver towards fuller extent of networking services is the case of cloud deployment, which ideally requires free allocation of network and computing functions over a number of servers and pods. This is difficult to arrange in fronthaul with point-to-point fibres and optical layer alone. One fronthaul-specific protocol stack difference from backhaul and midhaul is that the data plane can also be Ethernet without an IP layer, while all other traffic sources are otherwise IP protocol based. A converged network layer for backhaul and fronthaul that utilizes the same nodes needs to deliver both Ethernet and IP services.

Taking into account these requirements and the technologies at hand from Chapter 8, one example of a converged network is the use of MP-BGP control plane-based VPN solutions which are used to provide Ethernet services as an overlay for fronthaul and IP VPNs as an overlay for midhaul and backhaul, with a common MP-BGP control plane and IP (/ MPLS) data plane underlay as in Figure 9.22.

Traffic flows from CPRI-based RUs are carried after adaptation to Ethernet (RoE) or after conversion to eCPRI. Ethernet flows are transported by Ethernet VPN services and IP flows of midhaul and backhaul by IP VPNs, including specific VPNs for network slicing. Once the common underlay is deployed, the IP phase of fronthaul is supported by provisioning IP VPNs instead of (or in addition to) the Ethernet service. Both services can continue to exist.

QoS is guaranteed for time-sensitive flows according to the requirements of the TSN fronthaul profile. The scheduling mechanisms from TSN are expected to be supportable in many cases with IP nodes also (routers, multilayer switches). TSN fronthaul defines strict priority scheduling as a mandatory rqeuirement, together with guaranteeing resources for time-sensitive flows. This needs to be supported for the fronthaul Ethernet service also in the case of converged fronthaul and backhaul. Explicit paths and reservations are

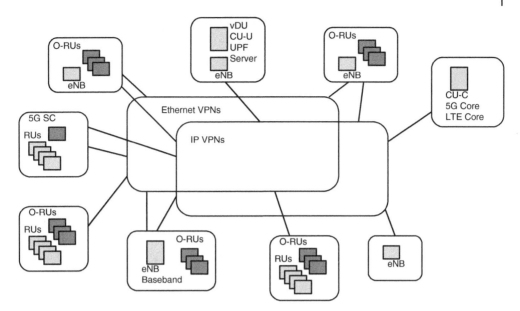

Figure 9.22 Merged network.

configured either by a central controller (management plane, centralized configuration) or by distributed protocols (distributed model) or by a hybrid format.

The network construct of Figure 9.22 fits well for solving cloud connectivity requirements. For pooling, dynamic allocation of resources and scalability, it is essential that network functions can be freely located within the servers and between data centres. A service overlay can be built on top of a common data plane architecture, which is based on MPLS or SRv6, for example, with MP-BGP managing the control plane for services like IP and Ethernet VPNs.

9.4 Indoor Deployment

Most of the mobile data is consumed indoors. 5G brings extreme broadband, but due to the vast range of frequencies allocated to 5G, different frequency bands penetrate in a different way into buildings. Also, buildings have new challenges as metal coatings have been introduced to windows to improve their heat insulation – these windows can attenuate radio signals to the level that it may sometimes be challenging to have good coverage indoors [7].

Lower frequencies penetrate better indoors than higher frequencies. For example, at 3.5 GHz, a window can introduce significant attenuation into the radio signal. It can still provide a good solution for home connectivity using, for example, home routers which have better antennas than typical handheld devices and provide connectivity to the end user via Wi-Fi, which most home devices and appliances support anyway. In areas where it is not economical to install fibre to the home, this can be an attractive alternative and provide much better connectivity for home users than xDSL (digital subscriber line) or cable modem. Capacity demands at home have increased, due to remote work requirements for video conferencing services, increased use of uplink data connection and increased utilization of streaming services for entertainment, for example.

When using mmWave, the attenuation through windows and walls for higher frequencies is much higher and this in practice means either a dedicated indoor solution for 5G or, in the residential use case, an external antenna outside the building to connect to the radio network and internal Wi-Fi access points that in turn provide indoor coverage. mmWave indoor deployments have started from locations such as stadiums, convention centres and airports, where throughput demand is high and a lot of people are in a small open area. mmWave indoor systems need to be planned so that the antenna is in sight of the user to provide extremely high broadband connnection.

Enterprise use cases have more strict requirements for coverage, capacity and latency than consumers. In factories, this means building dedicated 5G indoor solutions. There is a lot of potential to modernize manufacturing by removing dedicated cabling between machines and the systems that are controlling and operating those machines and assembly lines. For example, for a manufacturing line that needs adjustment and changes, the rewiring of robots may take weeks. If control is wireless, such changes require less manual work. But before you can replace a wireline system with wireless connectivity, you need to build good radio coverage – most likely a private indoor 5G system which can be implemented using a frequency dedicated for enterprise or leased from an operator. For data security purposes, it is likely that enterprises will increasingly install their own wireless systems, where the data does not leave the enterprise premises, but the whole network is owned by the enterprise.

Earlier indoor systems, especially for 2G and 3G, were built using a Distributed Antenna System (DAS). Such systems were often shared between different operators, especially in public buildings, and provided in the best case a good coverage solution. These legacy indoor systems are difficult to expand, and the RF cabling introduces high attenuation for 5G frequencies. Therefore, it is almost mandatory to have a separate indoor system for 5G.

The technical solutions for dedicated 5G indoor systems are based on typically the same baseband solution as macro, at least in the beginning of 5G. There are a few different solutions:

1) Indoor gNB typically consisting of smaller RRHs with integrated antennas that connect with a standardized twisted pair cable for CPRI/eCPRI over Ethernet and power over Ethernet to a hub that is in turn connected with fibre to the baseband unit.
2) The baseband unit can connect with hybrid fibre to the indoor RRH. Hybrid fibre is used for describing a cable that has fibre connectivity for data and copper connection for power.
3) Fibre connectivity to RRH that has integrated antennas or external antennas and local power source. A local power source is needed when the RRH is of higher power and therefore its power consumption is too high to be provided by other means.

The above technical solutions are illustrated in Figure 9.23. The figure does not indicate if there is a need for a separate baseband solution per technology, or a separate/combined

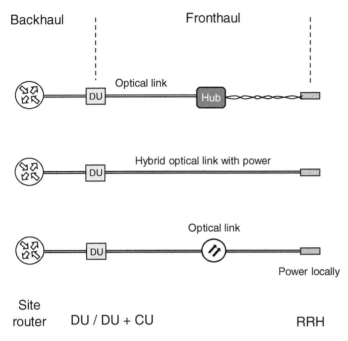

Figure 9.23 Technical solutions for indoor coverage and capacity.

antenna for each technology – it is just a simplified illustration for different connectivity options. The gNB can be either a combined CU + DU in the same location, or the CU can be located for example in the (far-) edge cloud. In the diagrams of this chapter, the DU can therefore be either DU or CU + DU.

In a typical office environment, it is easy to deploy an indoor system as described in option 1. Ethernet cabling is used to connect the antennas and Power over Ethernet (PoE) is used to provide power to the RRH. The Ethernet cabling already available in the building can be reused if it is category 6A or better. Indoor RRHs in most cases have integrated antennas for ease of installation, but external antennas are an option for special deployment scenarios. The hub is placed in the Telco closet where the switches and routers are located in such buildings, and its main purpose is to convert fibre connection to electrical and provide power to the RRH. The baseband unit is typically in the basement or at the top of the same building or a building close by, typically not more than a few kilometres away. This depends on the operator deployment model, if the baseband is centralized or distributed.

Similar systems have been deployed already with WCDMA and LTE, and 5G is added on top in different ways, depending on the vendor. There are low-power RRHs that provide connectivity for multiple radio access technologies (multi-RAT) or single radio access technology RRHs that are optimized for a single technology. The indoor solution can be built only for 5G, assuming existing DAS or outside macro to provide connectivity for other technologies. The baseband can be dedicated for each technology and combined for antenna connectivity at the hub or combined base solutions as possible.

In large buildings there might not be a suitable Telco closet within reach of 100 m from the planned antenna or RRH location, which is the typical reach for Ethernet cabling. Even if extenders can be used to extend the Ethernet and power over longer than 100 m distances, these solutions typically only reach up to 200 m. For distances longer than 200 m, fibre connectivity needs to be used and power needs to be provided – either locally to the RRH or using a hybrid cable that provides optical cable for connectivity and a DC solution for power (options 2 and 3 in Figure 9.23).

In a single operator solution, backhaul connectivity is provided to the baseband unit and the connectivity to the RRH provides fronthaul connectivity. For enterprise customers, large buildings and campus areas, this is an easy and fully controlled deployment model for operators. It is also the preferred model for operators, as it makes their corporate customers stick to their services.

Sharing the indoor solution in public buildings and shopping centres is often required, as building owners do not want multiple antennas installed by different operators and might require rent or fees that make the business case challenging. DAS systems have been easy in terms of management and responsibility, as different operators manage their own baseband units and backhaul equipment and connect them to the shared passive antenna system. Using active indoor solutions, there could be separate operators or neutral hosts that manage the indoor system and provide service to the operators, or one operator can be the lead operator and offer others connectivity to their system.

Synchronization is provided to the gNB in the baseband unit either by the transport network using G.8275.1 or G.8275.2, or locally using a GNSS solution. Synchronization is carried further to the RRH using the CPRI or eCPRI connection. In the NSA case, LTE is also recommended to have phase synchronization to allow best 5G and 4G interworking. When the connection from the baseband to the antenna is converted to eCPRI, synchronization also needs to be carried all the way to the RRH – the eCPRI does not include any specification for the synchronization.

The baseband unit of the gNB is connected with IPsec to the operator network in cases where the backhaul security is mandated either by the regulator or by the operator. Other operators may consider that restricting physical access to equipment is enough security, and backhaul connection is not required to be secured using IPsec. The fronthaul connection does not provide direct connectivity to the operator network, and interfering with this connection will result in a loss of connection between the baseband unit and the RRH.

An example indoor deployment is illustrated in Figure 9.24 where the baseband unit (indicated with DU) and the site router providing the backhaul connection are located in the basement of a building and then the hub connecting the baseband with fibre and providing Ethernet connection to the RRHs is located close to where the coverage is needed (i.e. the floors).

In factories and places where the environmental conditions are more challenging due to dust, temperature or coverage needs, it is practical to use lower-power micro RRHs that could also be used outdoors. These connect with fibre to the baseband unit and power is provided typically to the RRH directly. These micro RRHs have either integrated antennas, antennas directly attached to the RRH or external antennas depending on the use case and installation environment. The same type of RRHs can then be installed indoors and cover outdoor locations too in places like ports, airports, factories and logistics centres. This type of installation is illustrated in Figure 9.25.

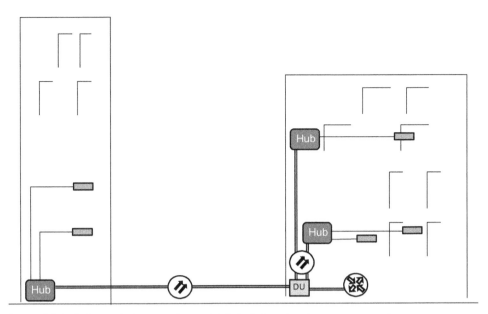

Figure 9.24 Indoor solution covering multiple buildings (e.g. in a campus area).

Especially with mmWave 5G, the capacity requirement in fronthaul connectivity is high and it would make sense to introduce a baseband solution closer to the RRH to save fibre connectivity. This will probably be the next development phase for 5G small cells, since deploying a macro baseband solution for smaller indoor locations is too expensive. This so-called 'all-in-one' small cell, as shown in Figure 9.26, would integrate baseband processing, radio and antenna, and could be connected with a high-capacity backhaul connection to the operator network. This would, however, require that synchronization is provided to every small cell using either a good-quality backhaul connection or separate GNSS solution.

The introduction of eCPRI will also decrease the fronthaul capacity requirement, or in other words the introduction of eCPRI will increase the fronthaul capacity and allow packet network usage in fronthaul. However, packet-based fronthaul considers only the user data and, for example, synchronization is outside of the eCPRI specification and needs to be provided to the radio unit or remote radio head separately using PTP. Synchronization over eCPRI introduces packet delay variation requirements for the eCPRI connection that need to be taken into account.

Cloud technologies will introduce disaggregation in mobile networks, as the software is separated from the cloud infrastructure and the hardware. In 5G, the baseband unit can be divided into virtual DU and virtual CU, and a lot of the processing required can be done on a commercial off-the-shelf computing platform, typically with L1 acceleration on top. In the future, this could also make it easier to share indoor infrastructure between operators when the radio processing is done in vDU and vCU applications running over generic cloud infrastructure and hardware. This will require that the vDU and vCU become cloud-agnostic applications, allowing them to be installed as software-only products.

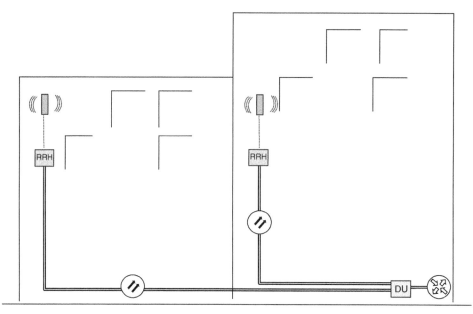

Figure 9.25 Micro RRH deployment in a building where higher output power in the RRH is required due to the large area and high ceiling.

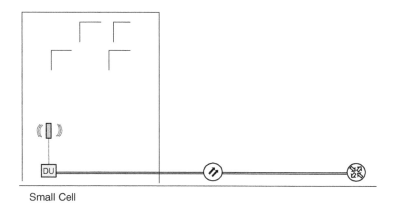

Small Cell

Figure 9.26 All-in-one small cell with backhaul connection.

9.5 Deploying URLLC and Enterprise Networks

9.5.1 Private 5G Examples

9.5.1.1 5G in Manufacturing

A smart factory is a typical example of the fourth industrial revolution, and there are multiple 5G use cases in factories. Some examples are mobile connectivity for Autonomous Mobile Robots (AMRs), production quality control based on video analytics and 'data showers'.

AMRs are used for automated material feeds, and they use mobile connection to communicate with an AMR fleet management system. While AMRs often include integrated local intelligence for navigation and safety, they can also send in uplink a large amount of sensor data and maintain a low-latency safety protocol connection with the central control system. 5G technology with URLLC enables an evolution to move part of the AMR computing resources from the AMR into edge servers in the network. Thus, 5G URLLC enables a more cost-efficient AMR solution with less computing resources in the AMRs. The 5G system must support high capacity, especially in uplink and low latency in radio, as well as in transport to send real-time data from LiDAR sensors for example and get an instant response for obstacle detection, automatic route changes and safe stopping to avoid collisions.

Uplink real-time video streaming is a common requirement for private 5G systems and manufacturing is no exception. Modern factories experience frequent (e.g. weekly) changes in factory layout. Flexibility of layout changes is maximized by minimizing the need for cables. Therefore, wireless connectivity is needed for production lines even if there is stationary equipment. Video analytics provide automated production quality control and 5G connected video cameras can be flexibly moved to locations as required, while analytics servers perform their tasks in factory IT cabinets.

Especially car manufacturing has a requirement for wireless 'data showers'. Modern cars have large amounts of data and software, which must be uploaded to in manufacturing. Furthermore, several automated software tests must be executed. This 'data shower' process should be executed as fast as possible wirelessly. This is possible with 5G extreme mobile broadband data connections with Gbps-range data rates.

9.5.1.2 5G in Underground Mining

Mining is one of the most advanced industries in automation. Autonomous haulage and tele-remote mining machines are mature technologies and have been developed using 4G and Wi-Fi. 5G improves the mobile connectivity and enables further innovations in mining solutions. 5G provides higher uplink capacity with reliable mobility for several video cameras, LiDAR and other sensors, for example from autonomous loaders in underground mines.

Automation generally improves productivity and in the mining industry improves safety significantly, as people can avoid entering dangerous areas. Sometimes it is adequate to support tele-remote control based on multiple uplink video feeds from mining machines and real-time remote control from a safe control room. However, fully automated solutions can operate without continuous active human control. Still, the mining solutions must provide video feeds and sensor data to control centres – including the option for human control in exceptional cases.

Autonomous underground mining benefits from automatic mapping tools, which create and update maps based on a high volume of collected sensor data. Real-time, high-quality video is also needed for visual operations monitoring. 5G technology has the capacity for a massive amount of uplink data generated by autonomous mining loaders and other equipment. There is also an increasing need to improve the collaboration between mining equipment, such as autonomous trucks and loaders. A 5G system with vehicle-to-anything communication enables flexible collaboration and automatic communication control.

9.5.1.3 5G in Technical Maintenance Services

Technical maintenance services for heavy machines and transportation vehicles need reliable wireless connectivity in working areas with a lot of metal around. Advanced 5G radio technology with features such as beamforming can provide high-performance connectivity for such maintenance services. Modern machine services can also include remote expert support, which in practice means that maintenance customers can participate, for example, in machine component inspection remotely and agree during inspection what type of maintenance is required. This is enabled with high-quality, real-time video connection from maintenance services to remote customer experts.

The 5G solution at maintenance service must have reliable coverage with a high uplink capacity for live video sharing, which can easily be implemented with 5G smartphones supporting for example 4K video quality.

9.5.2 Private 5G RAN Architecture Evolution

Standardized 5G RAN enables flexible placement of RAN functions and use of cloud servers in enterprise data centres. The transport requirements and solution vary depending on the placement of RAN functions. The architecture and deployment of RAN functions can also evolve over time. The presented examples show only standalone non-public networks, focusing mainly on RAN evolution to cloud-based RAN solutions.

The first private 5G networks use so-called 'classical' 5G base stations (or gNodeBs), where all 5G NR functions run on dedicated hardware units (see Figure 9.27). The gNodeBs connect over an IP backhaul network to 5G core functions running in general-purpose server hardware. Platform options for 5G core functions are bare metal servers, virtual machines or a cloud-native model with 5G core functions running in containers. Regardless of the 5G core platform, the core functions are usually close to other enterprise servers (i.e. in an enterprise data centre or other room with IT racks). When the 5G core is in enterprise premises instead of the operator central core, it is usually referred to as being in the 'edge cloud' or even 'far-edge cloud', meaning it is geographically very close to the RAN.

The first step towards RAN cloudification is to place the CU (i.e. non-real-time higher layer 2 and layer 3 radio functions) as general-purpose servers. With virtualization, CU

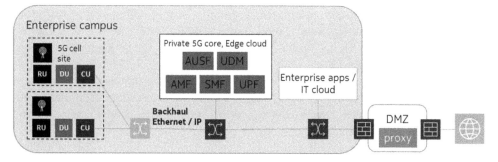

Figure 9.27 Small/medium campus network – classical 5G RAN (RU, DU, CU on dedicated HW components).

functions may share a common HW platform with 5G core functions. When the CU is deployed at a central site from cell sites, the transmission network between the DU and the CU is called midhaul (Figure 9.28).

The next evolution step is to move the DU functions to edge cloud server racks (Figure 9.29). In this case the basic IT servers are not suitable, but optimized servers with HW accelerators are needed. It should be noted that this is not extraordinary nowadays, because various modern applications like video analytics need HW acceleration GPUs. The fronthaul transmission between DU and RUs can be based on dedicated transmission links or use high-capacity Ethernet switches. Ethernet CPRI standards and implementations enable fronthaul transmission with regular high-end Ethernet switches (Figure 9.30).

Large multi-site enterprise networks can utilize enterprise WAN connections to place 5G core functions at centralized data centres (Figure 9.31). URLLC within a campus site is guaranteed by having the campus edge cloud at least with the core UPF, as well as end-to-end URLLC slicing.

Cloud RAN evolution in large enterprises (as in Figure 9.31) is similar to RAN evolution in small networks, as depicted in Figures 9.27–9.30. URLLC as well as high overall RAN performance require that RAN cloud functions stay local in each campus (i.e. in the local edge cloud).

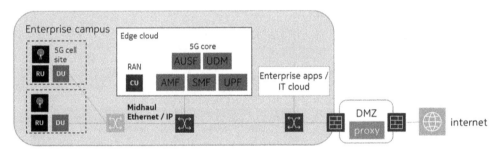

Figure 9.28 Small/medium campus network – distributed 5G cloud RAN.

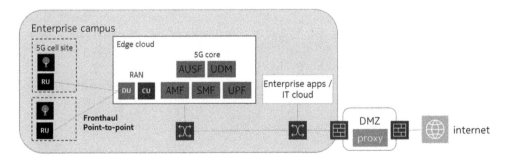

Figure 9.29 Small/medium campus network – centralized 5G cloud RAN with point-to-point fronthaul.

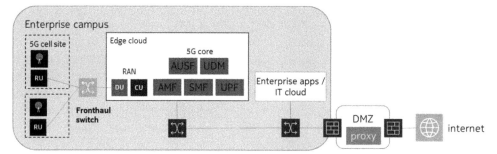

Figure 9.30 Small/medium campus network – centralized 5G cloud RAN with fronthaul switch.

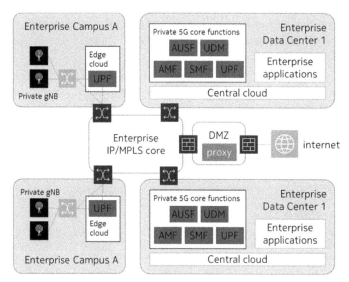

Figure 9.31 Large enterprise network – classical 5G RAN.

9.5.3 IP Backhaul and Midhaul Options for Private 5G

The first private 5G networks are based on classical gNB solutions. The gNBs are connected with IP/Ethernet backhaul to the 5G core in the edge cloud at enterprise premises. Technically, the backhaul solution for a private network can use the enterprise infra or it can be fully dedicated to the private 5G network. Typical backhaul solution options and their characteristics are described in Table 9.1.

Cloud RAN evolution requires IP connectivity for the midhaul between gNB CU and DU functions. The transport requirements for midhaul are similar to backhaul requirements.

9.5.4 Fronthaul for Private 5G

5G RUs support either CPRI or eCPRI fronthaul links, while eCPRI is expected to dominate in new RU products. Fronthaul is commonly based on point-to-point 10G or 25G (dark) fibre links. If the fronthaul distances are long, it is possible to use switches and provide a resilient fronthaul network. Various switching technologies can be used, such as optical

Table 9.1 Typical backhaul options.

Backhaul model	Integration with enterprise infra	Considerations
Fully dedicated IP/ Ethernet network (cabling, switches, routers) for 5G backhaul	• UPF N6 integration to enterprise IP infra	• IP plan, dimensioning and monitoring of backhaul does not depend on other enterprise infra and traffic
Virtual L2 overlay on top of enterprise infra (e.g. EVPN, VPLS or E-PIPE over MPLS)	• UPF N6 integration to enterprise IP for service • L2 service for Xn, N2 and N3 (gNB, AMF, UPF)	• IP addressing for 5G functions is independent of enterprise's IP address management • QoS and dimensioning of L2 service is required for backhaul traffic
L3 connective provided by enterprise infra	• UPF N6 integration to enterprise IP for service • Enterprise allocates IP addresses for Xn, N2 and N3 interfaces	• 5G functions are part of enterprise IP network • QoS and dimensioning is required for backhaul traffic

DWDM switches and Ethernet packet switches. Even with switches, the logical topology is point-to-point from DU to RUs (e.g. 25G E-Pipe to each RU).

9.5.5 Other Transport Aspects in Private 5G Networks

9.5.5.1 Media and Dimensioning

10 Gbps Ethernet is the starting point for connecting classical gNB backhaul. RAN cloudification and centralizing DU and CU functions can require even higher-capacity Ethernet ports for midhaul and backhaul. IP forwarding with routers is supported in midhaul and backhaul networks. Fronthaul ports are at minimum 10G ports, but 25G is common also. Modern eCPRI is based on Ethernet framing and therefore Ethernet switches can be used in fronthaul networks.

9.5.5.2 QoS

Fronthaul links have very tight latency and jitter requirements and therefore fronthaul is typically implemented with dark fibre or as logical point-to-point links with guaranteed capacity. eCPRI takes care of internal prioritization between control, user, synchronization and management planes.

Backhaul and midhaul connections are commonly connected over switched and routed networks. It is even possible that the IP/Ethernet network is shared with other enterprise traffic. Thus, QoS support is critical for midhaul and backhaul traffic. QoS mechanisms are needed for traffic marking, queueing and scheduling for E2E QoS differentiation (e.g. for eMBB and URLLC applications). Just as with eMBB services, QoS in transport is needed for proper E2E QoS differentiation between different applications. For example, private 5G networks can carry voice traffic, industrial control and safety protocols and high-bandwidth video streams simultaneously. These can be mapped to different 5G QoS flows mapped to different GBR or non-GBR 5QIs, which can be further mapped in the transport network to different IP and Ethernet QoS classes.

9.5.5.3 Synchronization

Accurate synchronization is mandatory for gNBs. It can be provided by connecting GPS receivers directly to gNBs. If GPS-based synchronization is difficult to arrange (e.g. for indoor or underground 5G networks), synchronization can be provided with the IEEE 1588v2 Precision Time Protocol (PTP) from a Grand Master Clock (GMC). The PTP must be prioritized over other traffic for accurate and well-performing synchronization. As described earlier, other 5G system elements may also require tight synchronization from a common GMC if the 5G system acts as a TSN bridge.

9.5.5.4 Security Aspects

Private 5G networks provide secure authentication, encryption and integrity over the radio interface. However, user data is not necessarily encrypted unless the enterprise has a complete E2E IP traffic encryption solution. If midhaul and backhaul traffic must pass the enterprise's firewalls, the firewall rules must be modified to allow 3GPP protocols.

9.5.5.5 MTU Size

Typically, enterprise IP networks support a 'standard' 1500-byte IP MTU for IP hosts and any IP applications. Similarly, L2 connectivity supports Ethernet MTU, which matches with a 1500-byte IP payload. When such enterprise application data is carried in midhaul and backhaul over the GTP-U protocol, the underlying transport network must support a larger IP MTU in order to avoid IP fragmentation. IP fragmentation would cause serious performance issues – especially for very high 5G bit rates – and E2E latency would increase. Fragmentation can also cause other issues such as packet drops in firewalls.

9.6 Delivering Synchronization

The 5G synchronization target as presented in Chapter 4 is at the Over The Air (OTA) interface, which is basically the antenna transmission point in the simplistic view. This means it is critical to ensure that the antenna site element is within the 5G system to achieve high synchronization accuracy matching the OTA interface synchronization target. For other 5G system elements away from the antenna site (such as 5G protocol stack higher-layer processing at the central site), a lower time-alignment accuracy for data transmission is tolerable. This relaxation is important, especially in the case of using Ethernet transport to separate the antenna site from other 5G system elements.

In this section, synchronization technologies to ensure the antenna site meets the OTA interface synchronization target are discussed. Currently, two technologies are considered suitable frontrunners:

- network timing synchronization using PTP and SyncE
- satellite-based synchronization

Note that this section primarily focuses on the technology description and comparison. It will treat the antenna site as a black box to support synchronization technology. The exact 5G system element at the antenna site will not be addressed, since this can be a complex topic based on 3GPP NR functional split and BTS deployment type (classical BTS vs centralized RAN vs cloud RAN).

9.6.1 Network Timing Synchronization Using PTP and SyncE

In general, it is quite complex to deploy the network timing synchronization solution. Many factors must be considered to ensure the synchronization target accuracy goal is met at the end user (i.e. antenna transmission point in the simplistic view). Specifically, we focus on the network timing synchronization solution using IEEE 1588 (Precision Time Protocol) and SyncE (Synchronous Ethernet) in the 4G/5G fronthaul network. The following specifications are the first-level standards that cover the fundamental specifications for using PTP to address the synchronization need in a telecom network:

- IEEE 1588–2008 (aka IEEE 1588 V2 or PTP). Standard for a precision clock synchronization protocol for networked measurement and control systems.
- ITU-T G.826x. Standards related to frequency synchronization with SyncE.
- ITU-T G.827x. Standards related to time and phase synchronization for telecom networks (PTP with optional SyncE).

The following specifications (relevant synchronization chapters) are the next-level standards that address the synchronization view in the fronthaul network for 4G and 5G deployment:

- IEEE 802.1CM. Time sensitive networking for fronthaul.
- IEEE 1914.1. Standard for packet-based fronthaul transport networks.
- ORAN-WG4 CUS. O-RAN Fronthaul Working Group – control, user and synchronization plan specification.
- CPRI TWG eCPRI. Common public radio interface – eCPRI interface specification.

Across all these standards, network timing distribution (via PTP and SyncE) is viewed as the primary solution to deliver synchronization across the fronthaul network. In general, these next-level standards provide an overall view – such as target synchronization accuracy goals for different 4G and 5G applications, and general network timing distribution topology/deployment models using the clock types defined in G.827x standards. For further details, such as analysis of the clock chain performance (i.e. number of specific clock types in a clock chain to satisfy a specific target accuracy goal), the common approach is to refer to the fundamental standards. Together with the fundamental level and next-level specifications, a comprehensive guide to deliver synchronization within the fronthaul network for 4G and 5G applications is possible.

9.6.2 SyncE

Refer to ITU-T G.8262 (EEC) and G.8262.1 (eEEC). SyncE alone is not sufficient to satisfy 4G and 5G RAN application synchronization needs. SyncE can only achieve frequency synchronization. However, PTP + optional SyncE can achieve time and phase synchronization, which is critical in 4G and 5G RAN applications. SyncE is optional, but will become mandatory when higher target time and phase synchronization accuracy goals are desired.

For example, with a fronthaul network that requires the use of higher-accuracy clock nodes (e.g. Telecom Boundary Clock, T-BC class C), syncE support in the network becomes mandatory. A clock node acting as T-BC (for PTP) to recover time and phase can simultaneously act as EEC/eEEC (for SyncE) to recover frequency. The hybrid PTP + SyncE support will lead to higher accuracy.

9.6.3 IEEE 1588 (aka PTP)

At the higher level, IEEE 1588–2008 (also known as IEEE 1588 V2) is a network timing solution based on a PTP. It is capable of achieving frequency synchronization and even phase/time synchronization between a master clock and a slave clock with high precision. It can be thought of as a good alternative between the older network timing solution (NTP, Network Timing Protocol) and the satellite-based synchronization solution. Compared to the NTP solution, it can provide higher accuracy with a similar network timing clock delivery approach. Compared to the satellite-based synchronization solution, it can achieve reasonably high and sufficient precision (even if it is not as good as the satellite-based synchronization solution) without the cost of GPS/GNSS receiver and antenna installation at each end-user site. It can also solve the deployment issue where GPS/GNSS signal access is difficult. The IEEE 1588 synchronization solution can eventually recover the time traceable to many international time standards, including GPS time. The GPS time epoch (starting at midnight January 5/6 1980) is typically used in RAN applications.

To understand the overall IEEE 1588 synchronization solution target for RAN applications, the IEEE 1588–2008 standard alone is not sufficient. Since the publication of IEEE 1588–2008, additional work has been done by the ITU-T organization over the years to create a set of recommendations on top of the IEEE 1588 standard to govern the overall solution. Relevant ITU-T specifications will be discussed in the next section.

For evolution, a newer version of the IEEE 1588 standard (IEEE 1588 V2.1) was published in 2020. The new version addresses many protocol and behaviour topics. Specifically for 5G synchronization for RAN applications, the goal to enhance support of synchronization towards 1 ns (instead of the sub-microsecond range in the current version) and to add optional provisions for protocol security are welcome enhancements. The PTP integrated security mechanism provides optional security for PTP. The PTP security extension provides source authentication, message integrity and replay attack protection for PTP messages within a PTP domain. It is also backward compatible with IEEE 1588–2008 to allow a mix of network elements with different IEEE 1588 versions. This is an important factor for any realistic and practical deployment scenario.

9.6.4 ITU-T Profiles for Telecom Industry Using SyncE and PTP

The complete set of standards and recommendations allow realization of a practical network solution supporting different network element interoperability. The following are the important standards and recommendations related to IEEE 1588 synchronization (with optional SyncE support) to use in the telecom industry, especially for 4G and 5G RAN applications:

- ITU-T G.826x. Using IEEE 1588 to achieve frequency synchronization.
- ITU-T G.827x. Using IEEE 1588 to achieve phase/time synchronization.

Specifically for 4G and 5G synchronization, we focus on the following recommendations:

- G.8271. Time and phase synchronization aspects of telecommunication networks.

For full timing support (network limits, clocks, profile):

- G.8271.1. Network limits for time synchronization in packet network with full timing support.
- G.8272. Timing characteristics of Primary Reference Time Clock (PRTC).
- G.8272.1. Timing characteristics of Enhanced Primary Reference Time Clock (ePRTC).
- G.8273.2. Timing characteristics of Telecom Boundary Clock (T-BC) and Telecom Slave Clock (T-TSC).
- G.8273.3. Timing characteristics of Telecom Transparent Clock (T-TC).
- G.8275.1. Precision time protocol telecom profile for phase/time synchronization with full timing support from the network.

For partial timing support (network limits, clocks, profile):

- G.8271.2. Network limits for time synchronization in packet network with partial timing support from the network.
- G.8273.4. Timing characteristics of T-BC-P, T-BC-A, T-TSC-P and T-TSC-A.
- G.8275.2. Precision time protocol telecom profile for phase/time synchronization with partial timing support from the network.

In general, reaching a high-accuracy time synchronization target (Category A, B or C as stated in eCPRI/802.1CM/ORAN/1914.1) with an IEEE 1588 solution is not trivial. Careful network planning must be followed. The recommendation based on full timing support (refer to ITU-T G.8275.1 profile) is universally agreed by all the guiding standards – as stated in Chapter 4 – to reach such a target. This means that all network switching elements must support the Telcom boundary clock (governed by ITU-T G.8273.2) or the Telcom transparent clock (governed by ITU-T G.8273.3) at a minimum. In addition, according to ITU-T G.8273.2, when the network also supports SyncE under ITU-T G.8262/8262.1, it is desirable to use both IEEE 1588 and SyncE together to speed up the start-up time and achieve target accuracy with more confidence. In the FTS profile, encapsulation is based on Ethernet framing with multicast only.

For a midhaul or backhaul network, partial timing support (refer to ITU-T G.8275.2 profile) is a possible option to provide timing towards CU and DU. In the PTS profile, encapsulation is based on UDP/IPV4 or UDP/IPV6 framing with unicast only.

9.6.5 Example of Putting All Standards Together in Planning

This section shows the logical steps to utilize the relevant standards to plan for a deployment. A simplified example is used to illustrate the concept. There could be a more complicated scenario in the actual planning, but that is beyond the scope of this section.

In the high level, planning consists of three basic steps:

1) timing topology and model selection
2) evaluation against target goal (Category A, B or C)
3) fine-tuning consideration

9.6.5.1 Timing Topology and Model Selection

In this example, the ORAN C3 configuration is used. This configuration plans to deploy PRTC/T-GM in the fronthaul network to drive all DUs and RUs. The G.8275.1 full timing profile is used. Switches as T-BC (or T-TC) will be used within the fronthaul network to connect DUs and RUs. A group of RUs can be used to support coordinate features that require relative |TE| accuracy to fulfil the Category A or B target goal. All RUs require absolute |TE| accuracy to fulfil the Category C target goal.

Based on the ORAN C3 timing distribution configuration concept (see Figure 9.32 for an example), the analysis can be based on the simplified model in Figure 9.33. RU1 and RU2 can be O-RU in the same remote site or across different remote sites, as shown in Figure 9.32.

Using this simplified model, the goal is to analyse two key parameters on any RU or RU pair.

- **Absolute |TE| to any RU (e.g. RU1 or RU2)**

This parameter addresses RU absolute timing alignment relative to an absolute timing reference (e.g. GNSS). This absolute |TE| parameter is important to evaluate against the Category C target defined in eCPRI/802.1CM/1914.1 standards.

Based on Figure 9.32 as an example:

$$\text{Absolute}\left|\text{TE}\right| \text{ to RU1} = \text{PRTC}/\text{T-GM error} + (r+1+n)\text{T-BC switch error}$$
$$+ \text{RU1 T-TSC error} \tag{1}$$

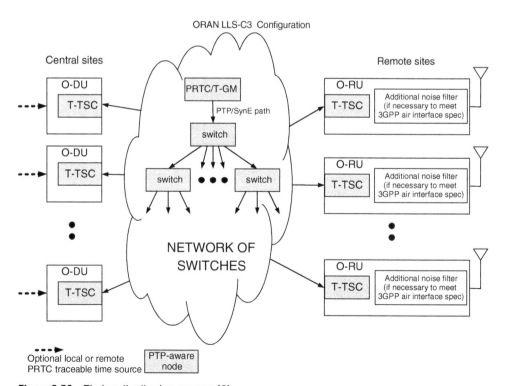

Figure 9.32 Timing distribution concept [8].

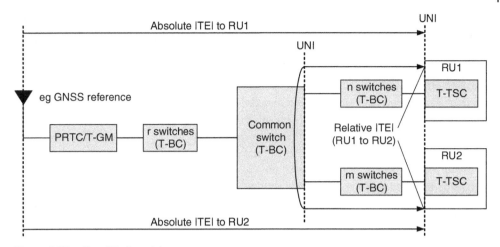

Figure 9.33 Simplified model.

$$\text{Absolute } |\text{TE}| \text{ to RU2} = \text{PRTC}/\text{T-GM error} + (r+1+m)\text{T-BC switch error}$$
$$+ \text{RU2 T-TSC error} \tag{2}$$

- **Relative |TE| between any RU pair (e.g. RU1 and RU2)**

This parameter addresses the RU relative timing alignment between any RU pair. This relative |TE| parameter is important to evaluate against the Category A or B target defined in eCPRI/802.1CM/1914.1 standards.

Based on Figure 9.32 as an example:

$$\text{Relative } |\text{TE}| \text{ between RU1 and RU2} = \text{common T-BC port } 1-2 \text{ error} + (m+n)$$
$$\text{T-BC switch error} + \text{RU1 T-TSC error} \tag{3}$$
$$+ \text{RU2 T-TSC error}$$

9.6.5.2 Evaluation against Category A, B and C

In all relevant standards (eCPRI/802.1CM/1914.1/ORAN), the target goal is to meet Category A, B or C by the end application (inside RU). The evaluation essentially identifies the number of hops (in terms of T-BC or T-TC) that can be allowed to fulfil Category A, B or C based on nodal specification of the clock types being used in a clock chain. The high-level evaluation concept is as follows.

- **Category C: ±1.5 μs**

To meet Category C, the RU recovered timebase (from T-TSC) must meet the timing error against an absolute timing reference (e.g. GNSS < ±1.5 μs).

In the above example, Equation (1) can be used to evaluate RU1 absolute |TE| for this category. Based on the nodal specification of each node (PRTC-/T-GM, T-BC in switch,

T-TSC in RU1), the maximum allowed number of hops of switches (r and n) can be determined. Similarly, Equation (2) can be used to evaluate RU2 absolute |TE| for this category to determine the maximum allowed number of hops of switches (r and m).

- **Category B (relative 260 ns) and Category A (relative 130 ns)**

To meet Category B or A, RU pair recovered timebases (from their corresponding T-TSC) must meet the relative timing error between the RU pair (< 260 ns or 130 ns).

In the above example, Equation (3) can be used to evaluate the relative |TE| between RU1 and RU2 for this category. Based on the nodal specification of each node (T-BC in switch, T-TSC in RU1 and in RU2), the maximum allowed number of hops of switches (m and n) can be determined.

The evaluation concept illustrated here is just an overview to give the reader the basic idea. For complete evaluation steps and proper summing formulas (RMS or static summing), the reader should refer to ITU-T G.8271.1. In the current ITU-T G.8271.1 standard, only the target goal of Category C is addressed. In the upcoming revision, it is planned to expand the scope of the analysis. The expanded scope should address the target goals of Category A and B with different clock types (e.g. T-BC/T-TSC class B or C).

9.6.5.3 Fine-Tuning Considerations

After steps 1 and 2 are identified and understood, the network can be planned with the following fine-tuning steps:

- Placement of T-GMs logistics
  - Deploy fewer T-GMs (e.g. one active GM plus one standby GM) to cover a larger group of RUs with a longer clock chain if possible. This strategy should be practical to meet the Category C target goal.
- Determination of maximum number of allowed clock node hops
  - Based on the intended nodal specification on switches (T-BC/T-TC class B or C) and RU (T-TSC class B or C), evaluate the maximum number of hops (i.e. T-BC switches) to meet the absolute |TE| for the Category C target goal.
  - Evaluate the maximum number of hops (i.e. T-BC/T-TC switches) to meet the relative |TE| for the Category A/B target goal.

- Placement of common switch for Category A/B
  - Place a common switch to feed an island of RU for the Category A/B target goal based on the step 2 limit.
  - If the step 2 limit is over, a possible adjustment could be to improve the nodal specification clock type (i.e. using a class C node vs a class B node) within that island.
  - If the step 2 limit is still over, even with the most accurate clock type (i.e. all class C nodes), a possible adjustment could be to deploy common nodes with more ports to feed the RU group with fewer intermediate switch hops. This will essentially shorten the clock chain.

- Placement adjustment for Category C or other absolute |TE| feature
  - If the number of clock node hops is too large to meet Category C or another absolute |TE| feature (such as OTDOA), even with the most accurate clock type (i.e. class C), a possible adjustment could be planning more T-GMs in the fronthaul network to shorten the clock chain affecting the RUs.

Note: Currently, there is a feature such as OTDOA that demands much tighter absolute |TE| than Category C. The requirement could be in the region of ±100 ns vs the Category C ±1.5 μs. Certain markets may request such a feature. A possible approach to meet this feature target is to deploy more T-GMs to shorten the clock chain. For example, a T-GM could drive an RU group to support OTDOA among these RUs. Across the boundary of different RU groups, the network operator would either prohibit OTDOA operation across the boundary or accept a degraded OTDOA performance only under the boundary condition.

9.6.6 Resilience Considerations in Network Timing Synchronization

One of the advantages of network timing synchronization is to support natural resilience. Based on IEEE-1588 PTP and SyncE standards, the protocol itself has provisions to support resilience as a deployment option.

For PTP:

- Multiple grandmasters (PRTC/T-GM) can be provisioned in a PTP timing network (domain) to support the active/standby redundancy concept. For simplicity, an example of one active grandmaster with several standby grandmasters is used to serve the entire PTP timing network. A more complex deployment to designate multiple active grandmasters within a PTP timing network (domain) to serve different end-application nodes is also possible.
- When an active PRTC/T-GM is down, a standby PRTC/T-GM can take over, and all end-application nodes (as PTP slaves) can relock to the new active PRTC/T-GM through PTP management communication. Careful network planning is needed to allow end-application nodes reaching different PRTC/T-GMs in order to support such a redundancy scheme.
- Besides PRTC/T-GM redundancy, there can also be intermediate clock distribution node (T-BC or T-TC) redundancy. Careful network planning is needed to allow each end-application node to reach the PRTC/T-GM via a different routing path, bypassing the faulty intermediate clock distribution node.
- The PTP standard (supporting Best Master Clock Algorithm, BMCA) and ITU-T G.8275.1 standard (supporting Alternate Best Master Clock Algorithm, Alternate BMCA) provide a mechanism to help establish the master–slave hierarchy within a PTP network. In the event of grandmaster or intermediate clock distribution node outage, BMCA (or Alternate BMCA) will kick in to re-establish the new master–slave hierarchy. After the conclusion of the BMCA (or Alternate BMCA), a PTP slave should eventually find a new and best possible path to reach a new active grandmaster. In a more complex deployment with multiple active grandmasters, it is also possible that the BMCA (or Alternate BMCA) conclusion is to establish a new path towards another active grandmaster.

For SyncE:

- Multiple masters (PRC) can be provisioned in the SyncE network to support the active/standby redundancy concept as well. However, it is less critical in a fronthaul network since SyncE support is mainly for hybrid PTP + SyncE mode at T-BC or T-TSC.

- In normal PTP + SyncE mode operation, the key requirement is to have a stable SyncE frequency over a relatively short interval for the PTP recovery algorithm to collect sufficient PTP timing packets. Hence, SyncE in holdover (within 4.6 ppm) is also sufficient to satisfy hybrid PTP + SyncE mode requirements. With this more tolerable performance goal, supporting SyncE redundancy or just SyncE holdover without redundancy is acceptable.

9.6.7 QoS Considerations in Network Timing Synchronization

When network timing synchronization is supported, PTP timing packet flow should be given higher priority to minimize the packet delay variation effect. The PTP standard relies on the underlying transport standards to support the high-priority objective.

- In the FTS profile, the PTP timing packet is based on transport using Ethernet encapsulation with multicast. The Ethernet frame header with VLAN tag (IEEE 802.1Q) supports 3-bit priority level based on IEEE 802.1P.
- In the PTS profile, the PTP timing packet is based on transport using UDP/IP encapsulation with unicast. The IP packet header supports EF (Expedited Forwarding) in the DSCP field based on RFC 3246 to support priority traffic through the IP network.

In fronthaul (transport between DU and RU), the focus is on the FTS profile, which is the mandatory profile to ensure RUs meet the stringent 3GPP air interface time alignment goal.

9.6.8 Special Considerations in Cloud RAN Deployment

In cloud RAN deployment, CU and part of the DU functionality that is non-real-time can be virtualized with a Virtual Machine (VM) at the central location. Real-time DU functionality is realized with dedicated hardware at the central location, or at the radio site with the RU. In either case, the network timing synchronization concept described in the previous sections can be used to ensure real-time DU and RU are synchronized with high accuracy.

In the cloud RAN scenario, non-real-time CU and DU functionality does not need accurate synchronization. In principle, it is sufficient for the non-real-time cloud portion to feed data into the real-time DU functionality in advance. For example, the non-real-time cloud portion can buffer data at a few milliseconds advance level and feed data into a jitter buffer in real-time DU functionality. Inaccurate synchronization effects at the non-real-time cloud portion, as well as any transport delay variation effect, can be absorbed by the jitter buffer. In addition, a feedback loop based on jitter buffer status can be used to guide non-real-time cloud portion transmission. Essentially, such a feedback loop can adjust the non-real-time cloud portion transmission time (slightly faster or slower) to keep the jitter buffer from overflowing or underflowing and avoid any transmission interruptions. Such a feedback loop can be implemented based on a specific design solution between the non-real-time cloud portion and the real-time DU portion.

9.6.9 Satellite-Based Synchronization

Besides the network timing synchronization solution as described in Section 7.9.1 to support fronthaul networks for 4G and 5G RAN applications, there is also the alternative to use (or continuously re-use) a satellite-based synchronization solution. In fact, prior to the arrival of the network timing synchronization solution, satellite-based synchronization (GPS or GNSS) has been used extensively for years as the traditional synchronization method in RAN applications.

In general, satellite-based synchronization generally refers to the use of a Global Positioning Satellite (GPS) system to acquire absolute time synchronization. The GPS system is a US-based satellite system with 32 satellites in the GPS constellation; it is well maintained by the US government. The GPS time epoch (starting at midnight January 5/6 1980) is typically used in RAN applications.

In recent years, satellite-based synchronization generally refers to the use of a Global Navigation Satellite System (GNSS), which expands the satellite system beyond just the GPS system. GNSS includes the United States' GPS system, Russia's GLONASS system, the European Union's Galileo system and China's Beidou system. GPS and GLONASS systems have been in operation for many years. The Galileo and Beidou systems are newer systems, with continuous expansion (more satellites being launched). The GPS time epoch is also available in GNSS, so it is a smooth transition for any end-user system moving from GPS-only to GNSS.

The GNSS receiver is used to receive GNSS signals from available satellites. In general, the GNSS receiver is versatile to support multiple (and simultaneous) system reception with the proper GNSS antenna. The ability to pick up more satellites from one system or multiple systems provides higher availability and accuracy. Supporting multiple satellite systems also eases any concerns over intentional shut down or degradation under government control.

In general, time recovery accuracy up to 15 ns (1σ) can be achieved when sufficient satellites are locked. Such high accuracy can easily meet the 5G synchronization target (mandatory ± 1.5 μs accuracy or down to a few 100 ns accuracy for advanced features). Compared to IEEE 1588 (with optional SyncE) synchronization, an obvious advantage is no constraint on the transport network for any synchronization upgrade (such as IEEE 1588 with optional SyncE-aware hardware). However, satellite synchronization also comes with a few inherent disadvantages:

- Satellite antenna installation complexity to ensure good visibility. This installation guideline can be particularly hard to meet in urban/indoor deployment.
- Intermittent (short-term) satellite loss. Satellite reception can be degraded due to weather conditions and/or sub-optimal installation. End-user equipment usually needs to provide failover capability to ride out this type of outage. In general, failover capability could be switching to a secondary back-up reference (e.g. IEEE 1588 network timing) or holdover with a local stable oscillator. In either case, there is an additional cost to support any failover scheme.
- Local jamming. The commercial use of GNSS does not provide additional security for the satellite link. Therefore, it could be subject to intentional disruption via local jamming. If this concern cannot be tolerated, an additional cost to support any failover capability becomes necessary.

9.6.10 Conclusion for Synchronization

The 5G system is still at an early phase and it is reasonable to expect many different types of deployment scenario. There will be classical BTS, centralized RAN, cloud RAN, urban/sub-urban/rural/indoor sites, macrocell/small cell type, and so on. Given both network timing synchronization and satellite-based synchronization schemes can meet 5G synchronization targets with their own pros and cons, a mixture of both solutions could be seen in deployments. A practical and cost-effective deployment scenario could mostly be a network timing solution in the network, with selected difficult-to-reach RU sites using satellite timing solutions. Difficult-to-reach RU sites may exist due to too many hops to reach a common switch/T-GM, but are not worth the investment of installing an additional common switch or T-GM.

References

1 3GPP TS 37.340 Multi-connectivity.

2 3GPP TS 38.425 NG-RAN; NR user plane protocol.

3 https://networks.nokia.com/solutions/airscale-cloud-ran.

4 http://www.cpri.info.

5 IEEEP1914.3 Standard for Radio Over Ethernet Encapsulations and Mappings.

6 O-RAN.WG7.FHGW-HRD.0-v02.00 Hardware Reference Design Specification for Fronthaul Gateway.

7 Karttunen, A., Mökkönen, M., and Haneda, K. (2019). Investigation of 5G radio frequency signal losses of glazing structures.

8 O-RAN Control, User and Synchronization Plane Specification.

10

Conclusions and Path for the Future

Esa Metsälä and Juha Salmelin

10.1 5G Path for the Future

5G genuinely opens new ways in which a mobile network can be utilized to benefit mankind, serving users in different use cases more efficiently and more effectively. At mobile system level, use cases that were previously possible only with a separate network can gradually converge to use the mobile network, when new service types like capability for URLLC is introduced. First voice transferred from the fixed network to mobile, then mobile broadband to a large extent and 5G has the potential to transfer further services to the mobile side, aided by network slicing capability.

In many areas, 5G is an important building block and stepping stone for even further enhanced capabilities that can follow the path 5G has started, both on the radio side and on the networking side. This means, for the air interface, new frequency bands and utilizing massive MIMO and beamforming to exploit the latest radio technology. On the networking front, cloud nativeness and friendliness will likely be further optimized, allowing easier deployment of different types of clouds, public and private, large and small, with possibilities to combine related applications into the same cloud. For this, the networking side strives towards a fully flexible and extendable fabric with a wealth of services.

Specifically for mobile backhaul and fronthaul this is an interesting phase, as requirements are very demanding: strict time sensitiveness of fronthaul, implementation of network slices, URLLC demanding ultra reliability and low latency, containerized network functions in the cloud requiring great flexibility and dynamicity from mobile backaul and fronthaul, and for high-frequency bands a large number of cell sites integrated into the rest of the network. All these development areas have to support uncompromised security with high availability in the network domain. Security will not just be important, but is a prerequisite as for most – if not all – applications, the network has to be trusted for all services that are delivered through it. A successful business model calls further for automized operations that allow quick provisioning of new network connectivity services and minimized complexity in operation.

5G Backhaul and Fronthaul, First Edition. Edited by Esa Markus Metsälä and Juha T. T. Salmelin.
© 2023 John Wiley & Sons Ltd. Published 2023 by John Wiley & Sons Ltd.

10.2 Summary of Content

This book started with a brief introduction to backhaul and fronthaul in Chapter 1. In Chapter 2, we had an inside view into the 5G system and 5G radio network, and the new technologies and approaches 5G has introduced – including the already mentioned massive MIMO radio, URLLC service, network slicing and the disaggregation of the 5G radio network. Chapter 3 offered a 5G operator perspective into what considerations are most essential from either a technological or an economic standpoint in the 5G network, and then specifically in 5G backhaul and fronthaul.

Chapter 4 analysed and summarized the key requirements for 5G backhaul and fronthaul based on the input given in 3GPP and the two previous chapters. Capacity is obviously a key requirement for any transport link, and the first one that comes to mind. The other areas covered were latency, availability, reliability, security and synchronization, all of which had major enhancements and changes from previous generations. As backhaul and fronthaul are to a large extent an implementation aspect, setting clear requirements allows multiple types of solutions to be built according to specific cases.

Some of the 5G system topics merit further discussion of implications and importance for transport as well. Network slicing is a major new development in 5G, and it is that for transport as well. Another transport-related area is IAB, which can be a very effective tool providing connectivity in the 5G access tier. The other issues covered were NTN, URLLC transport and non-public networks for industrial applications, and the development of smart cities. All these have, or may have, a big impact on 5G backhaul and fronthaul – as outlined in Chapter 5.

Chapters 6 and 7 focused on discussing physical layers with optical and wireless links. Chapter 8 continued with cloud and networking technologies, such that the combination of Chapters 6, 7 and 8 could be used as a toolbox to deliver solutions for Chapter 9, which continued with specific deployment cases.

10.3 Evolutionary Views for Backhaul and Fronthaul

After a review of the discussions in this text, from Chapter 1 through to Chapter 9, it is worth briefly considering backhaul and fronthaul transport and technologies as they are today and discussing possible evolution therefrom.

One line of development is in the fronthaul area, where the evolution so far has led from CPRI/OBSAI links to eCPRI with Ethernet and IP. The low-layer split point itself may evolve further and can be further optimized based on targeted needs and performance. A network likely starts to have gradually more complex topologies where resilience, statistical multiplexing and resource pooling become important. Generally, IP can act as the underlying technology, on top of which Ethernet and IP services are delivered, as required – offering commonality with midhaul and backhaul network segments.

Definitely another area that is developing further is that of disaggregated RAN and related to that, cloud deployments with different approaches in building radio clouds. This introduces many items for 5G backhaul and fronthaul:

- the level of disaggregation dictates the logical interfaces and connectivity needed with latency requirements for those interfaces, and as a consequence, it defines the physical distances at which different elements can be located from the cell site
- with cloud nativeness, computing resources ideally should be freely and dynamically allocated from a pool, when needed, and so these resources need to be reachable flexibly
- network domain security to the peer endpoints needs to be arranged in cloud deployment while maintaining high availability and dynamicity, with a trend where security is moving closer to the application
- a combination of local UPF function, local server and different radio clouds can deliver specific services (e.g. for industrial applications with network slicing)

With data centres, other applications may be combined with the radio applications with local UPF and local application servers. Smaller geographical areas – factories, harbours and ports – may use 5G infrastructure and the transport links there may be more campus-area and tailored according to the specific case, possibly requiring TSN-type technologies. Enterprise and industrial applications with URLLC services have huge growth potential. From the backhaul and fronthaul side, these cases vary a lot and require specific solutions tailored to the case, however still relying on common networking technologies and principles for robustness, security, availability and ease of operation.

While many technologies can be deployed to deliver connectivity, IP MPLS VPN principles have grown even further in importance and the combination of BGP control plane (MP-BGP) and IP (e.g. SRv6) or MPLS data plane is providing a rich set of services that match the requirements for wide deployment, resilience, security and flexibility for both Ethernet and IP VPN connectivity. A unified data plane which can serve as an underlying layer is key for a converged transport network, on top of which services are provisioned as needed. On the networking side, key protocols like BGP have already been enhanced in many ways and this will likely continue. For cloud networking, flexible connectivity means also pooling gains and possibilities for cost reduction in the servers, for the number of active servers needed and for their power consumption.

The access tier in mobile backhaul and fronthaul appears to be growing in value, since more and more possible tenants could utilize the connectivity that is built for the cell sites. An example is smart cities. This and other needs make sharing of transport links more important, and there again many approaches can be used – from a neutral host to more traditional common transport sharing methods.

A big part of the overall picture lies in managing and operating backhaul and fronthaul efficiently. Multi-domain coordination is required, and software-defined networking principles with domain-specific controllers and a higher-layer orchestrator is in many cases essential. These tools help provide for the QoS configurations in the network domain, including for demanding cases where paths need to be set up and resources reserved – for URLLC network slicing and other cases. Cross-domain tools also enable dynamic provisioning of network services for 5G network slicing, covering not only transport or 3GPP mobile network domains, but both.

A related aspect is the increasing level of openness in the network, which helps the efficient operation need described above. Open network models defined by YANG and configurations via Netconf/Restconf reduce the complexities related to multi-vendor environments and cross-domain coordination – both of which are a reality in almost all mobile transport networks and at networking layers in general. Open networks, and open management models, are simultaneously driven by many organizations – O-RAN, IETF and IEEE, to name a few.

Openness in the network standards continues, with the retiring of legacy protocols and their replacement with standard networking technologies – in 5G transport, an example is CPRI/OBSAI moving to Ethernet and IP. This in turn allows converged networks, where the same principles can be leveraged, with the network domain delivering the services, itself also evolving continuously.

Index

Page numbers referring to figures are *italics* and those referring to tables in **bold**.

5G Backhaul and Fronthaul, First Edition. Edited by Esa Markus Metsälä and Juha T.T. Salmelin.
© 2023 John Wiley & Sons Ltd. Published 2023 by John Wiley & Sons Ltd.